LIGHTING REDESIGN FOR EXISTING BUILDINGS

LIGHTING REDESIGN FOR EXISTING BUILDINGS

Craig DiLouie, LC

THE FAIRMONT PRESS, INC.

CRC Press
Taylor & Francis Group

Library of Congress Cataloging-in-Publication Data

DiLouie, Craig, 1967-

Lighting Redesign for Existing Buildings / by Craig DiLouie. -- 1st ed.
    p. cm.
Includes index.
ISBN-10: 0-88173-571-X (alk. paper)
SBN-10: 0-88173-572-8 (electronic)
ISBN-13: 978-1-4200-8385-9 (taylor & francis distribution : alk. paper)
1. Electric lighting--Design. 2. Buildings--Repair and reconstruction. I. Title

TK4175.D56 2011
621.32--dc22

2010051299

Published by The Fairmont Press, Inc.
700 Indian Trail
Lilburn, GA 30047
tel: 770-925-9388; fax: 770-381-9865
http://www.fairmontpress.com

Distributed by Taylor & Francis Ltd.
6000 Broken Sound Parkway NW, Suite 300
Boca Raton, FL 33487, USA
E-mail: orders@crcpress.com

Distributed by Taylor & Francis Ltd.
23-25 Blades Court
Deodar Road
London SW15 2NU, UK
E-mail: uk.tandf@thomsonpublishingservices.co.uk

Printed in the United States of America
10 9 8 7 6 5 4 3 2 1

10: 0-88173-571-X (The Fairmont Press, Inc.)
13: 978-1-4200-8385-9 (Taylor & Francis Ltd.)

Table of Contents

Acknowledgments

Special thanks to Craig A. Bernecker, Ph.D., FIESNA, LC, founder of The Lighting Education Institute (LEI), for renewing this book to help ensure it achieved a high level of technical accuracy.

Introduction

There is no more worthy, more glorious or more potent work,
than to work with light.
—Omraam Mikhaël Aïvanhov, philosopher and alchemist

Because light is used to see but cannot be seen, it is often overlooked by its users. The presence of light is rarely noted, in fact, unless there is a perceived lack of it (dim atmosphere or shadows) or a perceived excess (glare). Like air, however, light is invisible and yet everywhere in the visual environment, as we cannot see without it. And seeing, in turn, is fundamental to economic and leisure activity for the large majority of the population who are sighted. Like electricity and clean water, light is an engine of progress.

Ubiquitous in the built environment, light is generally considered a commodity, and the main economic consideration is how to obtain it more cheaply. For this reason, educated owners of new and existing buildings are finding it highly profitable to invest in energy-efficient equipment to reduce lighting energy costs by as much as 50 percent or more. These projects often take the form of simple lamp and ballast replacements.

These investments can be great uses of capital. They can also be a waste of money, even if the project achieves an excellent financial return based on lower operating costs. The reason is simple: Light is for people, not buildings. Decisions about light may be all about dollars and cents, but dollars and cents are not all about energy.

In short, it is not enough for light to be energy-efficient. It must also be *effective*. What does it mean for a lighting system to be energy-effective?

First, we must understand that the application of light is lighting—not only the equipment that acts as the delivery system for light, but also where the equipment places light and with what direction, intensity and color. As people respond to varying levels of brightness and color in the field of view, light can be applied to the same space to impact a building and its occupants in different ways. While light makes sight, lighting is about perception—whether a space appears tense or uninteresting, public or private, spacious or intimate, productive or relaxing, and so on.

Some 80 percent of sighted people's impressions of the world, in fact, are generated by what they perceive with their eyes. This is where lighting delivers tangible economic benefits beyond simple vision. Properly applied, lighting can produce higher sales of key merchandise, optimize the productivity of office workers, offer a memorable experience for visitors, beautify space and architecture, improve learning rates, influence human interaction and mood and atmosphere, and promote safety and security.

It should be noted that bad lighting can realize opposite effects.

Given the benefits of good lighting design, and the high economic stakes involved, we may be thinking about light all wrong. Yes, light is a commodity. But *lighting* is a business asset—a critical asset of both the built environment and the organization that occupies it.

In existing buildings, this asset is often neglected. According to the Department of Energy's 2003 Commercial Buildings Energy Consumption Survey, while lighting upgrades are a popular renovation investment, lighting upgrades have been performed in only 29 percent of commercial building floor space built before 1980. This suggests that about 25 billion sq. ft. of floor space is still lighted to pre-1980 standard using T12 lamps, magnetic ballasts and overlighted spaces.

Upgrading these lighting systems to today's lighting efficiency standards could generate lighting energy cost savings of up to 50 percent or more, according to the National Lighting Bureau. As the fluorescent magnetic ballast becomes virtually eliminated in 2010 and most fluorescent T12 lamps with it in 2012, building owners should begin exploring opportunities to convert their lighting systems to more-efficient technologies in a way that achieves maximum benefit. But focusing solely on how much energy a lighting system uses is like buying a forklift based solely on its fuel efficiency instead of how much it can lift, how it handles, and its speed and safety features. After all, the purpose of lighting is not to draw wattage, but instead, depending on its application, to enhance task performance, provide visual comfort, reveal form and architecture, attract interest and so on. In short, lighting should be effective as well as efficient.

Here, too, research suggests that the lighting asset is often neglected in buildings in terms of lighting quality, perpetuating lighting systems that may have been poorly designed or designed to outdated standards, and poorly maintained since then. According to a 1999 office lighting survey conducted by office systems manufacturer Steelcase, 37 percent of workers said the lighting in their workspace was either too dim (22 percent) or too bright (15 percent), while three out of four said they wanted more control over light levels. Further, three out of four office workers said better lighting would improve their efficiency and productivity, while two out of three said they would be more creative.

This means simple replacement of lamps, ballasts and controls is not enough. A component-based retrofit approach may save energy, but perpetuate a poor design that fails to achieve the business goals of the organization that invested in owning the asset and wants to realize its value. Instead, the lighting system may need not retrofit, but relighting—a redesign that addresses effectiveness as well as energy performance, including issues such as visual comfort, uniformity, π and light on walls and ceilings. Because it is not enough for a lighting design to be efficient; it also has to shine.

Some of these issues run deep, and can be challenging to address properly at very low levels of energy consumption. As complexity increases due to advancing lighting technologies and an imperative to optimize lighting quality as well as energy efficiency, so has demand for greater expertise from designers of lighting systems.

This book was written to educate owners, energy managers, electrical engineers, architects, lighting designers, consultants, electrical contractors, electrical distributors and other interested professionals about the relighting of existing buildings. The information may apply to lighting design in new construction as well. Its thinking transcends my first book about lighting upgrades, *The Lighting Management Handbook*, published by The Fairmont Press more than 15 years ago, challenging owners and designers to optimize lighting quality hand in hand with efficiency in existing buildings—and get the full value of an asset that is effective as well as efficient.

I hope you find this book useful in designing with light. The challenges are daunting. The opportunities are compelling. Let us begin.

Chapter 1

Retrofit or Relight?

If you retrofit an inefficient, lousy lighting job with new lamps
and ballasts, the result is an efficient, lousy lighting job.
                              —Willard L. Warren, energy consultant

ADVANCED LIGHTING FOR SAVING ENERGY

Energy efficiency is America's cheapest energy source; it is less expensive to invest in using a unit of energy more efficiently than to generate it. As buildings, not vehicles, are the country's leading source of carbon emissions—a significant byproduct of electric generation as well as auto engine combustion—energy efficiency is also just as important as renewable energy to reduce air pollution and mitigate global warming. The U.S. Department of Energy (DOE) has established an aggressive goal of achieving net zero energy buildings—buildings that consume no net energy and produce no carbon emissions—by 2025, and considers energy efficiency to be part of the solution.

For owners and managers of commercial buildings, energy efficiency is potential opportunity residing in their building systems. If tapped, it is profit that increases every year that energy costs increase. If left untapped, it is a significant opportunity cost that increases every year that energy costs increase. The average cost per kilowatt-hour (kWh) in 2009 was $0.1021 for commercial buildings, about 40 percent higher than the average for the year 1999 (see **Table 1-1**). What's more, the average annual increase in energy costs has outpaced the average rate of inflation over the past 10 years while labor and equipment costs remained relatively stable. This means projects that may not have met organizational criteria for investment in energy efficiency as a cost avoidance strategy one year may qualify in later years (see **Figure 1-1**), meriting periodic evaluations.

Table 1-1. Average retail price of electricity to ultimate customers by end-use sector, by state, 2009, in cents per kWh. Source: Department of Energy, 2010.

State	Average Retail Price Commercial (cents/kWh)	Average Retail Price Industrial (c/kWh)	State	Average Retail Price Commercial (cents/kWh)	Average Retail Price Industrial (c/kWh)
Alaska	14.65	13.38	Montana	8.20	5.63
Alabama	10.00	6.05	North Carolina	7.92	5.93
Arkansas	7.62	5.85	North Dakota	6.79	5.90
Arizona	9.38	6.62	Nebraska	7.32	5.69
California	13.73	10.46	New Hampshire	14.74	13.50
Colorado	8.24	6.31	New Jersey	14.35	11.38
Connecticut	16.68	16.81	New Mexico	8.55	5.81
District of Columbia	13.91	10.15	Nevada	10.61	7.96
Delaware	11.96	9.30	New York	15.40	9.71
Florida	10.74	9.18	Ohio	9.59	6.69
Georgia	8.89	6.13	Oklahoma	6.90	4.94
Hawaii	21.86	18.14	Oregon	7.74	5.58
Iowa	7.45	5.17	Pennsylvania	9.56	7.17
Idaho	6.52	5.15	Rhode Island	13.63	12.79
Illinois	8.30	7.53	South Carolina	8.64	5.74
Indiana	8.16	5.72	South Dakota	7.04	5.66
Kansas	7.97	6.16	Tennessee	9.54	6.74
Kentucky	7.58	4.89	Texas	9.84	6.99
Louisiana	7.85	5.26	Utah	6.97	4.80
Massachusetts	17.80	11.61	Virginia	8.10	6.87
Maryland	11.98	9.91	Vermont	12.89	9.29
Maine	12.54	9.94	Washington	7.03	4.34
Michigan	9.61	7.17	Wisconsin	9.50	6.70
Minnesota	7.87	6.28	West Virginia	6.77	5.23
Missouri	6.88	5.32	Wyoming	7.29	4.84
Missisippi	9.51	6.58	U.S. Total	10.21	6.84

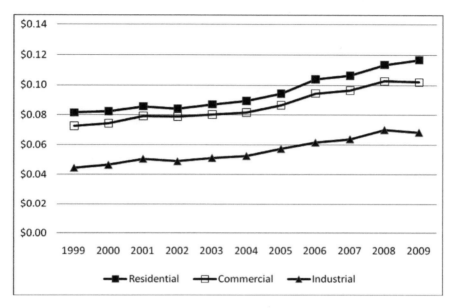

Figure 1-1. Average residential, commercial and industrial sector electric energy costs, 1999-2009. Source: Department of Energy, 2010.

Energy-saving Lighting Opportunities

Lighting is a major energy user in the United States, accounting for about 20 percent of total electric energy consumption, according to DOE. In 2006, commercial buildings spent $38.8 billion on electricity to operate lighting systems, or 32 percent of the total spent by commercial buildings in the country, making it the largest consumer of site electricity out of all building equipment, from HVAC to office equipment and PCs. The generation of this electric energy resulted in 263.1 million metric tons of carbon dioxide emissions, or 32 percent of the total produced by commercial buildings in the country.

Lighting systems consist of luminaires, lamps and any required auxiliary devices, and controls. The *luminaire* (commonly called a *light fixture* in the field) is a complete lighting device that includes a *light source*, auxiliary components such as *ballasts*, electrical connections, housing and an optical system directing light in a controlled direction and pattern. The light source, or *lamp*, is the actual light-producing component, generally classified according to method of converting electricity to illumination: *fluorescent, high-intensity discharge (HID), incandescent/halogen, light emit-*

ting diode (LED). Auxiliary devices such as ballasts provide sufficient voltage to ignite the lamp and regulate current during operation. Finally, *lighting controls* provide the ability to manually or automatically, based on an understood input signal, turn lighting ON or OFF or raise or lower light levels. Looking at the space as part of the lighting system, additional elements include room and object surfaces and daylight, if present.

In commercial buildings, advanced lighting and control systems are available that can replace existing systems for significant energy savings. On August 22, 2008, Secretary of Energy Samuel W. Bodman wrote an open letter to building owners inviting them to participate in a national effort to make buildings more energy-efficient, stating that "cost-effective lighting technologies are available now to cut energy costs by up to 50 percent." Based on 2006 energy expenditures—and assuming 25 percent of buildings have already been upgraded—this would cut the nation's electric bill by up to an estimated $15 billion—annual savings that would fall to the bottom line and make American industry more efficient and competitive.

Now let us get more specific for a moment. According to the *2009 Buildings Energy Data Book* published by DOE, the average large office building ($\geq$25,000 square feet, or sq. ft.) has a lighting power density (LPD) of 1.3-1.8 watts (W)/sq. ft. (based on 2006 data), while the average small office building has a lighting power density of 1.7-2.2W/sq.ft. If these buildings were relighted to simply comply with the ASHRAE 90.1-2004 energy standard, which imposes a maximum allowable LPD of 1W/sq.ft. using the Building Area Method, lighting power savings of 23-55 percent could be achieved.

Of course, this is just if we were to reduce LPD to code, which is supposed to be considered a minimum; even deeper savings are possible using paths outlined in tools such as the ASHRAE Advanced Energy Design Guides and the DOE Commercial Lighting Solutions webtool. These savings also do not count significant additional energy cost reductions that can be gained through the use of automatic lighting controls such as occupancy sensors (see **Table 1-2**). As the average large office building operates 16 hours/day—4,190 hours divided by 52 weeks divided by an assumed 5 days/week—such savings could be substantial, as shown in **Table 1-3**.

Energy data also indicates that the average school building built before 1980 has an LPD of 1.8W/sq.ft. while the average building built after 1980 has an LPD of 1.7W/sq.ft. If these buildings were relighted to

Table 1-2. Energy savings potential of popular advanced lighting control strategies in various building spaces.

Space Type	Controls Type	Lighting Energy Savings Demonstrated in Research or Estimated as Potential	Study Reference
Private Office	Occupancy sensor	38%	*An Analysis of the Energy and Cost Savings Potential of Occupancy Sensors for Commercial Lighting Systems*, Lighting Research Center/EPA, August 2000.
	Multilevel switching	22%	*Lighting Controls Effectiveness Assessment*, ADM Associates for Heschong Mahone Group, May 2002.
	Manual dimming	6-9%	*Occupant Use of Manual Lighting Controls in Private Offices*, IESNA Paper #34, Lighting Research Center.
	Daylight harvesting (sidelighting)	50% (manual blinds) to 70% (optimally used manual blinds or automatic shading system)	"Effect of interior design on the daylight availability in open plan offices", by Reinhart, CF, National Research Council of Canada, Internal Report NRCC-45374, 2002.
Open Office	Occupancy sensors	35%	National Research Council study on integrated lighting controls in open office, 2007.
	Multilevel switching	16%	*Lighting Controls Effectiveness Assessment*, ADM Associates for Heschong Mahone Group, May 2002.
	Daylight harvesting (sidelighting	40%	"Effect of interior design on the daylight availability in open plan offices", by Reinhart, CF, National Research Council of Canada, Internal Report NRCC-45374, 2002.
	Personal dimming control	11%	National Research Council study on integrated lighting controls in open office, 2007.
Classroom	Occupancy sensor	55%	*An Analysis of the Energy and Cost Savings Potential of Occupancy Sensors for Commercial Lighting Systems*, Lighting Research Center/EPA, August 2000.
	Multilevel switching	8%	*Lighting Controls Effectiveness Assessment*, ADM Associates for Heschong Mahone Group, May 2002.
	Daylight harvesting (sidelighting)	50%	*Sidelighting Photocontrols Field Study*, Heschong Mahone Group, 2003.

Table 1-3. Potential energy savings if average office, school, retail buildings and hospitals and medical facilities upgraded to LPD levels in the ASHRAE 90.1-2004 energy standard. Source for average building data is DOE.

Building	Average LPD	Average Operating Hrs/Yr	ASHRAE 90.1-2004 LPD Cap[1]	Energy Savings[2]
Office				
Large ($\geq$25,000 sq.ft.)	1.3-1.8W/sq.ft.	4190	1W/sq.ft.	23-44%
Small (<25,000 sq.ft.)	1.7-2.2W/sq.ft.	3340	1W/sq.ft.	41-55%
School				
Built Before 1980	1.8W/sq.ft.	2436	1.2W/sq.ft.	33%
Built After 1980	1.7W/sq.ft.	2436	1.2W/sq.ft.	29%
Retail				
Large ($\geq$25,000 sq.ft.)	1.6-2.1W/sq.ft.	4500-5245	1.5W/sq.ft.	6-29%
Small (<25,000 sq.ft.)	1.7-2.2W/sq.ft.	3786-4412	1.5W/sq.ft.	12-32%
Hospital	2.1W/sq.ft.	6752	1.2W/sq.ft.	43%

[1]Using the Building Area Method.
[2]Based on lighting power savings based lamps, ballasts and luminaires only, not including lighting controls. Even deeper savings are possible by using tools such as the Commercial Lighting Solutions webtool, ASHRAE Advanced Energy Design Guides, Collaborative for High Performance Schools (CHPS) guidelines, etc.

comply with ASHRAE 90.1-2004, which restricts school/university building LPD to 1.2W/sq.ft., lighting power savings of up to 33 percent could be achieved, not counting additional impacts from adoption of lighting controls. Even deeper energy savings are possible using the ASHRAE Advanced Energy Design Guides or Collaborative for High Performance Schools (CHPS) guidelines.

The average large retail building ($\geq$25,000 sq. ft.), meanwhile, has an LPD of 1.6-2.1W/sq.ft., while the average small retail building has an LPD of 1.7-2.2W/sq.ft. If these buildings were relighted to comply with ASHRAE 90.1-2004, which restricts retail building LPD to 1.5W/sq.ft., lighting power savings of 6-32 percent could be achieved. Again, this does not count lighting controls. Even deeper energy savings are possible using the ASHRAE Advanced Energy Design Guides or Commercial Lighting Solutions webtool. The Commercial Lighting Solutions webtool, for example, is designed to enable good lighting quality for up to 30 percent energy savings over and above ASHRAE 90.1-2004.

And regarding hospitals and medical facilities, the average building has an LPD of 2.1W/sq.ft. ASHRAE 90.1-2004 imposes a maximum allowable LPD of 1.2W/sq.ft. for hospitals and 1W/sq.ft. for healthcare clinics, suggesting an energy savings potential of 43 percent for hospitals and 52 percent for healthcare clinics. Not including controls, and again, even deeper energy savings are possible using the ASHRAE Advanced Energy Design Guides.

The most lucrative opportunities include lighting systems with one or more of these attributes: high energy costs, obsolete and relatively inefficient lamping and ballasting, inefficient or poor quality luminaire design, overlighting for the tasks, luminaires operating when they are not needed, excessive use of dark finishes that absorb light, long operating hours and poor maintenance practices. Examples of inefficient equipment being targeted by energy regulations include most fluorescent T12 lamps, magnetic ballasts for full-wattage and energy-saving 4- and 8-ft. T12 lamps, mercury vapor ballasts, probe-start metal halide ballasts in new luminaires, and incandescent reflector lamps.

Driven by energy concerns and energy codes, the market has already shifted to more-energy-efficient lamps, ballasts and controls in new construction, but less-efficient lighting is still consumed in U.S. commercial building floor space measured in billions of square feet. Consider that in 2005, 163 million T12 fluorescent rapid-start (mostly 4-ft.) lamps were sold in the United States; for every more-efficient T8 medium bi-pin fluorescent (mostly 4-ft.) lamp sold in 2005, about 0.75 T12 lamps were sold. In 2005, 27 percent of all fluorescent lamp ballasts sold in the U.S. were magnetic ballasts. And as recently as 2001, incandescent lamps were being used to provide illumination in 38.5 billion sq. ft. of commercial building floor space, or 62 percent of the total, not including mall buildings, according to DOE.

Lighting in Older Buildings

As these conditions are more likely to be encountered in older buildings, let us determine the potential for lighting upgrades in buildings built before 1980. According to the 2003 U.S. Energy Consumption Survey produced by DOE, there were about 4.6 million commercial buildings operating in the country, not including malls, representing about 64.8 billion square feet of floor space. The average nonresidential building was about 30 years old and operated about 50 hours per week.

About 2.5 million of these buildings, or 60 percent of the total, representing about 57 percent of total floor space, were built before 1980. According to DOE, about 988,000, or 39 percent, have benefited from some type of renovation, with lighting upgrades being a popular renovation second only to cosmetic improvements by number of buildings, and cosmetic improvements and HVAC equipment upgrades by quantity of floor space. And yet, looking at all pre-1980 commercial buildings—including those that have been renovated and those that have not—lighting upgrades have only occurred in about 449,000 buildings, or about one of five (18 percent) of the total built before 1980, as shown in **Figure 1-2**. Additionally, lighting upgrades have occurred in only 10.3 billion sq. ft. of building floor space, or less than one-third (29 percent) of the total built before 1980, as shown in **Figure 1-3**. Floor space built before 1980 is estimated at about 35 billion sq. ft., or 56 percent of total lighted commercial non-mall building floor space in 2003.

This suggests that about 25 billion sq. ft. of commercial building floor space may still be lighted to pre-1980 design standards (overlighted) and likely using the least-efficient lighting systems allowed by law. These spaces represent some of today's most lucrative candidates for energy-efficient lighting upgrades. Looking more deeply at the DOE data, we see that the most significant end-use markets are office and education buildings, and buildings in the Northeast and Midwest.

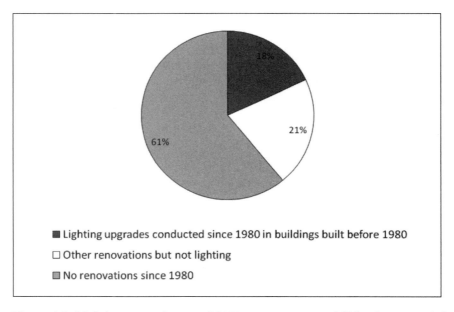

Figure 1-2. Lighting upgrades, as of 2003, as percentage of *lighted commercial non-mall U.S. buildings built <1980*.

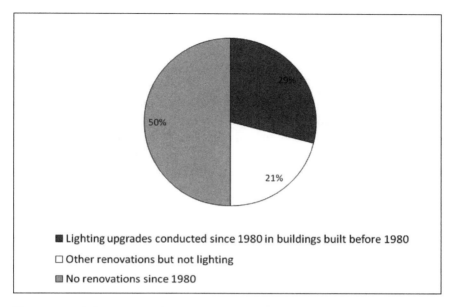

Figure 1-3. Lighting upgrades, as of 2003, as percentage of lighted commercial non-mall U.S. floor space built <1980.

Landlords and Tenants

One of the biggest barriers to lighting upgrades in office buildings has traditionally been who pays/who benefits conflicts between landlords and tenants in commercial lease properties. In the office market, this is a common problem, as tenants occupy nearly one-half of building floor space, according to DOE data. A typical high-rise building can include dozens, even hundreds, of leases, and many of them may address the subject of utility costs slightly differently. The lease may be a gross lease, in which utility costs are paid by the owner and calculated into the fixed rent; net lease, in which utility costs are passed through to tenants; and a fixed-base lease, in which utility costs are locked in over the term of the lease, with the owner paying for increases in utility costs or benefiting from decreases.

In today's business climate, landlords and tenants share an interest in saving energy. Landlords are willing to make deals and tenants are highly interested in cutting costs.

The building owner regards the building as an income-producing asset and stands to benefit significantly from investments in energy efficiency by reducing costs (if a gross or fixed-base lease), making lease spaces more marketable (if a net lease), and increasing the asset value of the property. Even if the lease is a net lease, it may be structured to enable the owner to recoup its investment before the energy savings are passed through to the tenant. In some situations, the owner may be able to split the savings with the tenant in exchange for an increase in rent. And at a 10 percent capitalization rate, a $1 decrease in bottom line expenses translates to a $10 increase in the present net value of the building.

The tenant, meanwhile, often regards energy efficiency upgrades as benefiting only the owner of the building, even though the remaining period of its lease may be much longer than the typical payback for energy-efficient lighting. If the lease is structured so that the tenant has to directly bear utility cost increases, the tenant may regard an investment in new lighting as more attractive, or put pressure on the landlord to reduce rent or utility costs or risk losing the tenant.

Lighting Upgrades in Newer Buildings

Good opportunities to generate energy cost savings through better lighting exist in buildings built in the last 10-20 years as well.

For example, in high-bay applications such as warehouses and school gyms, high-bay fluorescent luminaires can replace probe-start metal halide luminaires—made obsolete themselves in 2009 by the Energy Independence and Security Act—for up to 50 percent energy savings. Prior to 2009, a building could be built with low-initial-cost metal halide lighting, and instantly become a candidate for 50 percent lighting energy cost savings by converting that installation to more-efficient fluorescent hi-bay luminaires.

Another candidate for energy-saving upgrades in newer buildings, which might be surprising to some, is older T8 lighting systems, such as 4-ft. basic-grade 700 series lamps operated by 0.87 ballast factor generic electronic ballasts, which can be upgraded with newer, more-efficient T8 systems. (In review, ballast factor is a light loss factor that describes the amount of light a given lamp and ballast combination produces; BF x initial rated lamp output = lamp/ballast system output.) Options include energy-saving T8 lamps in 23W, 25W, 28W and 30W models operated on low (0.71-0.78) and normal (0.87-0.88) ballast factor instant- or programmed-start ballasts. Some of these energy-saving T8 lamps are now dimmable. Other options include using high-lumen lamps with a low ballast factor ballasts or using a high-lumen system and then reducing its size through delamping: Examples include high-performance T8 ("Super T8") lamps with a low, normal or high (1.14-1.20) ballast factor ballast, or dimming ballast. Potential energy savings can be as high as 18-30 percent—or more if lower light levels are acceptable.

If the existing lighting system has reached a point at which the ballasts are starting to fail, consider a lamp-ballast system upgrade with expanded options for maximum energy savings. For example, NEMA Premium high-efficiency electronic ballasts, which can generate higher energy savings, can be a good source of savings. If the existing lighting system is still relatively new, consider replacing the standard 32W T8 lamps with energy-saving models of T8 lamps on the existing ballasts (if light levels allow), particularly if they are 4-ft. basic-grade 700/SP series T8 lamps rated at 2,800 lumens, which are being phased out by DOE regulation along with most T12 lamps starting July 14, 2012.

In each case, the higher the applicable energy rates or the more an existing space is overlighted, the better the payback and return on investment, and the greater the cost-saving opportunity for the owner.

ADVANCED LIGHTING FOR PEOPLE

As stated in the introduction, while saving energy is important, lighting's purpose is not to use electricity, but instead, depending on the application, to support task performance, provide visual comfort, reveal form and architecture, attract interest and so on. While light is a commodity to be obtained for the lowest cost, lighting is not: It is a business asset that must be effective for its owner to realize its value. Effective lighting can enhance vision and task performance, draw attention, influence social discourse, create atmosphere, beautify architecture and spaces and increase comfort. All of these positive effects can contribute to desired business outcomes such as greater sales, workforce satisfaction and so on.

Poor lighting can have the opposite effects, however, and often does: Lighting, a key workplace characteristic, often fails to do its job, according to a survey of office workers, who complained about glare, eyestrain and other symptoms of poor lighting. According to the 1999 Steelcase Workplace Index Survey, 86 percent of respondents said that making lighting improvements in their work spaces would reduce eyestrain and headaches. Eight out of 10 experienced glare, with reflected glare being reported on computer screens (57 percent) and reading material (37 per-

In the Spotlight: The Lighting Upgrade Process

1. **Ask,** "Why are we doing this?" to prioritize desired benefits of a lighting upgrade, including lighting quality, energy savings, improved space appearance, lower maintenance costs, improved safety and security, and so on.
2. **Gather education, training and resources** to ensure that a project team is in place with the right combination of knowledge and skills to evaluate lighting systems based on all project criteria.
3. **Assess** the existing lighting system to determine LPD.
4. Conduct an **initial determination** to determine if lighting improvements would produce desirable financial opportunities.
5. **Survey** the existing lighting system to determine the type and location of lighting equipment in addition to whether it is producing appropriate quantity and quality of light.
6. Conduct a **system design** to identify good lighting upgrade candidates.
7. Perform an **audit** comparing these candidates to the existing lighting system and identifying proposals that satisfy the lighting goals with the most attractive financial performance. Include the impact of financing opportunities such as performance contracts, tax incentives and utility incentives.
8. Write **specifications** for the proposed lighting system.
9. **Acceptance**: The audit and specifications are accepted and the project is ready to move forward.
10. Conduct a **pre-installation survey** of the existing space to identify installation issues such as scheduling, storage, obstacles, etc.
11. **Conduct a trial installation** for the proposed lighting in a prototype space to evaluate its performance (optional).
12. **Install** the new lighting and properly dispose of the old lighting, recycling or donating older equipment when possible.
13. **Commission** the lighting system, including aiming and adjusting luminaires as needed, and ensuring any installed lighting controls are calibrated, oriented and working together according to the design intent.
14. **Acceptance**: The project is inspected to ensure proper installation, all operating and maintenance manuals are turned over to the owner, any required training is scheduled and performed, and any required explanation of the new lighting to occupants is provided.
15. **Energy verification**: The lighting may be submetered to verify projected energy savings, benchmark consumption, conduct ongoing load analysis, etc.

cent), and direct glare being reported (18 percent). Nearly four out of 10 workers said the light level in their workplace was either too dim (22 percent) or too bright (15 percent). When work areas were perceived as too dim, people either brought in their own lighting from home (15 percent) or got it from their employer or coworkers (13 percent). When work areas were too bright, 15 percent of the office workers had either removed or blocked the offending source of glare—sometimes involving removing lamps or attaching cardboard to monitors. Three out of four wanted more control over their lighting. Nearly half wanted to be able to adjust light levels with a dimmer switch (48 percent), control the amount of glare (46 percent) and be able to move lights around (26 percent).

These are not lighting problems. They are productivity problems. Office workers relate their lighting conditions to their productivity and creativity levels. According to a study in *Journal of Applied Psychology*, workplace characteristics account for as much as a 31 percent variance in work satisfaction, which accounts for 63 percent of variance in organization commitment, which in turn accounts for 80 variance in intent to turnover. Obviously, this can detract from the bottom line: The Gensler 2006 U.S. Workplace survey, in fact, suggest that poor productivity due to workplace design costs corporations more than $300 billion each year. The good news: The Steelcase Workplace Index Survey found that 86 percent of respondents among surveyed office workers believed their mood and energy level would be improved if their lighting problems were solved, while 75 percent said it would improve their efficiency/productivity, and 66 percent said it would improve their creativity. It is easy to put a value on a man-hour; what value do you put on creativity, on ideas that can grow a business?

There are several important lessons to be learned from this. First, the most aggressive energy-saving projects increase risks to lighting quality that require good lighting expertise. It's a sign of the times: The demand for higher expertise among lighting designers has increased along with increasing complexity of lighting technologies and design requirements. Saving energy is easy; providing good lighting while maximizing energy savings is getting more challenging all the time. Consider the results of a survey on energy-efficient lighting conducted by this book's author in 2006: When asked to rate various drivers and barriers to energy-efficient lighting, architect, lighting designer and electrical engineer respondents rated "provides the greatest value to the client" as a major demand driver, but also cited "unacceptable tradeoffs with lighting quality" as a major demand barrier. One can conclude from this that respondents place a high value on energy-efficient lighting, but are concerned about potential quality compromises.

Several years ago, the Center for the Built Environment (CBE) conducted a web-based satisfaction survey of more than 34,000 occupants in 210+ buildings—including 15 LEED-certified and six "self-nominated" green buildings (seven of them award-winning projects)—that is relevant to the question of whether "efficient" also means "good." The study indicated that while LEED/green buildings received statistically significant higher scores from occupants for general building satisfaction, indoor air quality and thermal comfort, they did not for lighting. In fact, while LEED/green buildings promote features such as daylighting, the level of user satisfaction with lighting, according to the occupant surveys, was about on par with conventional buildings. The number of LEED/green buildings surveyed was insufficient for strong conclusions, but the results were suggestive.

Charlie Huizenga, Research Specialist for CBE, explained to this book's author at the time: "Twenty-five percent of the LEED/green buildings were in the top 10 percent of lighting satisfaction scores, while 25 percent were in the bottom five percent. In other words, there was a real split in the lighting performance—several LEED/green buildings performed very well, and several performed very poorly, resulting in a similar average to the conventional buildings."

He added: "Reading through the comments in the survey, we found that the commonly identified problems were low light levels, glare and poor controls. These are issues that need to be addressed in design."

This leads us to the next lesson, which is that lighting quality should be properly prioritized in the project. As lighting designer Naomi Miller once wrote, "Remember there are people behind LEED points." When there is a conflict between energy efficiency goals and lighting quality requirements, consider which has a greater impact on the bottom line. The risks and benefits can be challenging to assess because lighting quality is more difficult to measure than efficiency. But remember: While energy savings are more easily measurable, user satisfaction can have a much greater impact on the bottom line. Consider, for example, that worker salaries and benefits can run up to $318/sq.ft. per year in a typical commercial building, while energy costs run about $2.25/sq.ft. per year, according to Carnegie Mellon University's Center for Building Performance and Diagnostics. If energy costs can be reduced by 50 percent through various efficiency measures, including an aggressive lighting upgrade, this would result in $1.13/sq.ft. in annual cost savings that fall straight to the bottom line. But if productivity could be increased by 1 percent, that would be $3.18/sq.ft. per year in economic gain. If your next project is a school, how much would a 1 percent increase in learning rates be worth? If your next project is a retail building, how much would a 1 percent increase in sales be worth? If your next project is a hospital, how much would a 1 percent increase in doctor and nurse environmental satisfaction be worth? While there is no simple metric for predicting the economic value of lighting quality, it should be understood via basic common sense that lighting is designed to support organizational goals and its success or failure impacts the ability of the organization to meet these goals.

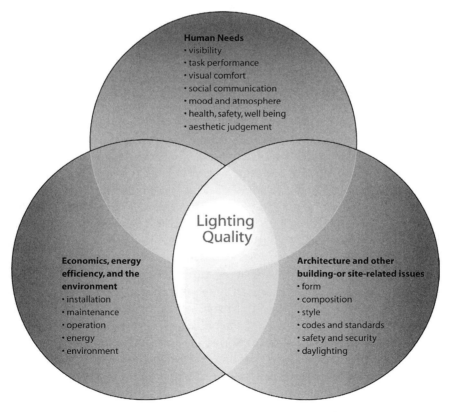

Figure 1-4. Energy efficiency is only a part of the story of providing a quality lighting design. Copyright 2010 Illuminating Engineering Society of North America; used with permission.

Now that we understand the value of energy efficiency and lighting quality in context of the needs of the given project, as well as the value of acquiring good lighting design expertise to ensure all requirements are met, the project's financial performance can be evaluated appropriately. This means that when considering a lighting upgrade, consider *upgrading the lighting*. In other words, evaluate and consider investing in improving the total performance of the lighting system, not just its energy performance. In some buildings, the lighting system is satisfying the organizational goals and an energy-saving project can be evaluated for investment on a purely financial basis as long as the new lighting follows the rule, "First, do no harm." In other buildings, the energy savings can be used as a basis of finance for a capital investment in new lighting that supports organization goals, such as worker satisfaction, retail sales, learning rates, etc.

Relight or Retrofit?

A key consideration for improving lighting in an existing building is whether to retrofit or redesign the lighting system.

In its *Guidelines for Upgrading Lighting Systems in Commercial and Institutional Spaces*, IES defines a *retrofit* as "modifying a lighting system so as to lower the operating costs or improve the performance of that system while not changing the original intent of the lighting design." Examples include lamp and ballast change-outs and direct replacement of reflectors, lenses, louvers, entire luminaires and controls. After a survey of the existing installation and proper measurement of its performance, retrofit proposals, which are shown in calculations to produce desired light levels, are vetted based on economic performance.

The *Guidelines* define a *redesign* as "modifying a lighting system so as to intentionally modify the goals of the existing lighting design." This basically entails starting fresh with some practical limitations, formulating an original design intent and installing a new lighting system that addresses concerns such as task visibility, emotional atmosphere, control systems, aesthetics, light patterns, ease of maintenance and other issues in addition to energy efficiency. A redesign may involve different luminaires, relocating existing luminaires, or producing a design with more of fewer luminaires. It may also be accompanied by interior design changes such as repainting room surfaces and replacing ceiling tiles to improve reflectance.

The first question for either approach is whether there is an opportunity to make changes to the lighting system to save energy. This can be addressed through a series of specific considerations.

Does the existing lighting system feature older technology such as incandescent lamps, probe-start metal halide systems, and/or fluorescent T12 lamps and magnetic ballasts? Are local utility costs (energy and demand charges) very high? Is the lighting uncontrolled and therefore operating longer than it is used each day? Was the building built before 1980 with lighting producing higher light levels than needed? Is the lighting system overall in poor condition and considered inadequate to continue serving its function?

These considerations are addressed during the assessment and initial determination phases of the upgrade process as defined by IES, during which a quick assessment is made of the building's LPD and the potential for lighting improvements is determined. Basically, the more "yes" answers there are to the above questions, the more likely a retrofit or redesign project will be an economical investment in energy savings. Newer buildings and recent renovations with short hours of operation are less likely to be cost-effective.

Retrofits and redesigns have their pros and cons as approaches. Some buildings may benefit from both approaches to address the needs of different spaces.

Retrofits are generally simple, minimally disruptive to operations, unobtrusive to the character of the space, and cost less than redesign projects, but may fail to realize lighting quality criteria considered important to achieving organizational goals. Additionally, altering existing luminaires by adding new components creates risks of matching interoperable components (such as T8 lamps on T12 ballasts) and nullifying manufacturer warranties. Energy savings result from replacement of inefficient equipment and subsequently reduce ongoing operating costs. In applications where lighting quality is not important or there is a major barrier to relighting, such as asbestos in the ceiling or prevailing wage laws creating a much higher cost for new luminaire installation, retrofits may be the only cost-effective option.

Redesigns can refresh the aesthetics of a space and more thoroughly address lighting quality needs as well as energy efficiency requirements, thereby potentially making a bigger impact on supporting organizational goals and building value. The older the lighting system, the more it would likely benefit from a redesign. Similarly, the worse the condition of the existing lighting—such as the school lighting shown in

Table 1-4—the more it would benefit from a redesign. Some redesigns take advantage of this approach's inherent flexibility and can generate much higher energy savings than typical retrofits, particularly when lighting control systems are applied to the solution. But they are typically more involved and costly processes, particularly if interior design changes are added to the renovation plan. In a redesign, energy savings result from optimizing the lighting system for efficiency and lighting quality. Building owners are apparently open to such capital improvements: According to a Light Right Consortium market study of professionals who specify, install and own/use lighting systems, 87 percent of respondents reported flexibility in lighting budgets if a return on investment could be demonstrated. This entails recognizing lighting's value beyond saving energy.

In some buildings, both approaches may be required. For example, an organization may determine that relighting the general task areas of a building is not cost effective, and subsequently retrofits the existing luminaires. In key areas, however, such as lobbies, executive offices, meeting spaces, computer-intensive classrooms, a redesign may be practical and cost effective.

Table 1-4. Percentage of public K-12 schools with "inadequate" building features, including lighting, 1999: Respondents at small, medium and large schools were asked to rate their public school facilities as "excellent," "good" and "adequate" ("adequate or better") or "fair," "poor" or "replace" ("less than adequate"). Small schools were defined as having 1-299 students, medium as having 300-599 students, and large as having 600 or more students. Source of data is the National Center for Education Statistics.

Building feature	Small	Medium	Large
Electric lighting	19%	17%	16%
Electric power	23%	21%	22%
HVAC	29%	32%	26%
Plumbing	28%	27%	20%
Interior finishes	20%	16%	18%
Exterior walls, finishes, windows and doors	31%	21%	23%
Framing, floors and foundations	19%	12%	14%
Roofs	24%	22%	22%

Figure 1-5. In new applications where the ceiling is not readily accessible to conventional general lighting solutions (as shown here), or in older existing redesign applications in which the ceiling plenum may contain asbestos, portable, furniture-integrated lighting may offer a possible solution. Photo courtesy of Tambient Lighting.

Have the building's primary spaces been retasked to new purposes for which the existing lighting system produces insufficient lighting conditions? Are existing light levels too low for the tasks being performed? Is uniformity poor, with uneven light levels and high contrasts between bright and dark areas—with one possible cause being the luminaires are spaced too far apart? Are the finishes of surfaces and objects dark and absorbing light? Do the primary spaces have poor lighting on walls and ceilings, making the spaces appear cavelike, or failing to provide adequate illumination for shelving? Are there points of interest that are not being properly highlighted using focal lighting?

"Yes" answers to these questions for each building space make these spaces candidates for a redesign. Another path to deciding whether a redesign is appropriate is if the retrofit options would have a negative impact on lighting quality.

Usually, the survey phase of the project—the initial building walk-through—will reveal to the trained eye the answers to questions related to both energy savings and lighting quality. Retrofits and redesigns are related; a retrofit is basically a lighting design with the luminaires already in place, while a redesign is a design with a relatively free choice of luminaire selection and placement. Many electrical professionals are good at thinking like accountants when they look at lighting systems; they may benefit if they think a little like lighting designers when they look at spaces. This means understanding not only what goes into a lighting system, but also what comes out of it and its relationship to people.

Acceptable lighting quality, according to the IES *Guidelines*, takes into account a number of factors that can be explored during a walk-through, including visual comfort, glare, uniformity, color rendering, lighting on walls and ceilings, and harsh patterns, shadows and flicker. These factors are important to thoroughly understand whether the project is retrofit or redesign, whether the goal is to improve lighting quality or simply maintain it.

This book advocates lighting quality as a critical companion to energy efficiency, not necessarily redesign as an approach. That being said, the remainder of this book, while entirely relevant to lighting retrofits, is more focused on principles related to redesign. Further, as redesign is similar to design, these principles may also be useful to electrical professionals involved in developing design criteria for lighting new buildings.

THE BUILDING SURVEY: SCOUTING FOR OPPORTUNITY

At Toyota, executives have used a process called the Treasure Hunt to find, prioritize and implement opportunities to reduce energy costs. Toyota defines the Treasure Hunt as an internal energy auditing process in which the existing energy management organization, working with process engineers, operations and maintenance personnel, identifies energy reduction opportunities across the company's operations. These opportunities are then prioritized based on difficulty of implementation and payback; select projects are funded.

Treasure Hunts have enabled the company to remain continually engaged in a process of energy improvement. The reason the Treasure Hunt works at Toyota is because it has top management involvement and support.

In the Spotlight: Common Lighting Mistakes

In the May 2009 issue of Buildings Magazine, David R. Laybourn, director of marketing and sales for Lime Energy in Glendora, CA, offers some good advice in an article titled, "Prevent 17 Common Lighting Mistakes."

Ten of these costly mistakes include:

1. Working with the wrong people: Choose experienced partners.
2. Conflicts in the chain of command: Invite input from all affected parties but empower a single project manager to make decisions from beginning to end.
3. Ignoring occupants: Include the lighting system's ultimate users in the process.
4. Skimping on the audit: Conduct a thorough audit to avoid mistakes and realize full value of the lighting upgrade.
5. Focusing on price: Avoid substitutions and inexperienced service providers that can compromise lighting quality or energy savings.
6. No contingency planning: Consider "what ifs" in your planning.
7. Not testing thoroughly: Thoroughly evaluate all options before commitment.
8. Sloppy installation: Ensure the installation is properly planned and coordinated.
9. Using averages to calculate financial performance: Use localized data, such as the building's actual energy rates, to conduct your economic analysis.
10. Missing opportunity: "It's a big mistake to believe that installing new equipment to save energy is 'not in the budget,'" Laybourn writes. "That's like saying you can't afford to save money. The mere act of paying your electricity bill means there's cash waiting to work for you."

A lighting upgrade starts with a commitment and a vision. Commitment must come from senior management to support the upgrade process. The vision is the benefits of quality lighting, fit to organizational priorities, that include efficient use of energy and increased human performance. Senior-level commitment and a thorough grasp of the benefits will provide the momentum to start and finish the lighting upgrade.

Data Collection

Accurate and complete information can help the project team choose the most appropriate lighting upgrade option, eliminate wasteful lighting practices, fully realize the benefits of good lighting, and provide an energy-effective lighting system for its users.

This information should include:

- building plans if available;
- interviews and survey data with building managers and occupants;
- utility billing history;
- operating hours; and
- detailed schedule of lighting equipment and controls.

Area-by-area Survey

An area-by-area survey of the lighting system is an important step in data collection. The first step is to obtain a floor plan or reflected ceiling plan of the facility. If one is not available, create a floor-by-floor line drawing of the facility on graph paper. Sometimes, what is actually used is a combination of floor plan and drawing as the building may have been remodeled since the actual floor plans were drawn.

Each area (room, transition area or group of rooms) should be labeled by number and a generic description of the space. During the area-by-area survey, a sticker bearing the area's description and number or simply the number can be adhered near the door hinges or other discrete place to match the area numbers on the floor plan. This can help with analysis, installation and maintenance.

For each room, identity:

- space characteristics, including room dimensions, whether daylight is present and, if so, how it is controlled, operating schedule and estimated annual operating hours, and reflectances on ceilings and walls;
- number and type of existing luminaires, including nominal size, number of lamps and ballasts/luminaire, type of lamps, type of ballasts, type of shielding material, total input watts, and physical condition; and
- primary visual tasks performed in the space, light levels on these tasks (including whether the space is underlighted or overlighted), range of age of occupants and any lighting quality problems.

In many buildings, time can be saved by using the plans to identify similar spaces and surveying a sample of those spaces.

Determining Lighting Hours

Determining lighting operating hours is critical to estimating financial return and identifying opportunities for lighting controls. There are four tools that can be used to estimate lighting operating hours:

- interviews with building management staff;
- surveys/interviews with maintenance staff and occupants;
- data loggers connected to dedicated lighting circuits in sample spaces; and
- existing lighting controls—manual switches, occupancy sensors, schedule-based lighting control systems, etc.—as revealed by the area-by-area survey.

Occupancy Survey

An excellent source of information to tap before starting a lighting upgrade is the experience of the occupants of the space that is lighted. An occupant questionnaire can be a simple and effective tool to collect information about how occupants regard their lighting system. Each group of questionnaires can be correlated to the room description and number in the layout.

The occupant questionnaire should be tailored to the characteristics of the facility and the type of work performed there. Questions put to a factory worker, for example, will usually differ from those asked of an office worker. The responses can provide insight into problems with the lighting system and opportunities to improve its effectiveness. Note that the responses will be from people who know little about lighting; in fact, they rarely notice the lighting system unless it malfunctions in some way. For example, when an occupant writes that there is too much light in an area, they may mean they think it is too bright. This may be a problem of lighting quality (glare) instead of lighting quantity. Try to word question so that they do not confuse the reader and yet specific enough so that the answers are not misleading as well.

Survey Tools

The area-by-area survey usually yields the most important information. While it can be a relatively simple process, it can require extensive legwork. Successfully conducting the survey depends on using the right survey tools, techniques and forms.

Before beginning the survey, the right equipment should be assembled and the surveyor should become acquainted with its proper use. This equipment may include an accurate light meter, mechanical counter, data loggers, measuring tape or infrared measuring device, access equipment, camera and a notepad and forms (or laptop). There are also situations that may warrant the use of a watt meter to measure actual luminaire wattage.

A high-quality illuminance meter (photometer), also called a light meter, should be obtained. This handheld device measures the amount of light impacting the work plane and room surfaces. Besides the ability to measure light level in footcandles (or lux), it can be used to determine the ability of room surfaces to reflect light onto the workplace.

Although counting luminaires appears to be a simple task, in a large area it can be frustrating when the surveyor's attention is distracted and he or she loses count. The mechanical counter is a handheld device used to count luminaires. To count one luminaire and record the number, simply press a button on the counter, which automatically advances the count by one.

Connecting data loggers to sample dedicated lighting circuits

Figure 1-6. Typically, an average light level reading for the entire work area is performed before cleaning and relamping and then after, using a light meter. Photo courtesy of Colorado Lighting, Inc.

can provide short-term monitoring of lighting operating hours.

Surveying will require equipment to access the lighting system. Indoors, a step ladder should allow quick access to most luminaires. Outdoors and in industrial areas lighted by high-bay luminaires, aerial platforms may be required. Access equipment provides the ability to measure energy consumption and determine the condition of luminaires.

In the Spotlight: Using a Light Meter

Before taking any measurements from a system in which new fluorescent lamps have just been installed, allow 100 hours of operating time for the lamps to stabilize. To get a true reading, no outdoor light can be allowed to enter the area. Since most lighting management operations happen after normal work hours, this may not be a problem. A procedure for measuring the amount of light in a space is provided below for general educational purposes. For specific procedures, consult the light meter manufacturer.

1. Expose the light meter's sensor to the system's light for about 15 minutes. This allows the meter to stabilize. If the fluorescent lamps are just being turned ON, leave them ON for at least 10 minutes to allow them to stabilize.

2. Take the meter reading (horizontal tasks). To gain a horizontal footcandle reading, lay the meter down flat on its back on a task surface (workplace), with the sensor facing the ceiling. Be careful not to cast a shadow over or near the meter so as to avoid distorting the reading (unless a shadow is normally cast over the workstation by the worker doing their job). Also avoid wearing a white shirt as this will reflect some light and distort the reading. Stand back from the meter when taking the actual readings to avoid blocking light and/or reflecting light onto the meter. A remote detector (photocell) will often help here, particularly when recording illumination at high workplace heights.

3. Take the meter reading (vertical tasks). Vertical tasks include warehouse racks and vertical merchandise displays in stores. Position the light meter on its base, with the sensor facing the room, and take the reading. Again, do not let any shadows or reflected glare hit the meter or you will distort the reading. Note the height at which the measure-

ment is taking place, which may be determined in advance through understanding IES vertical illuminance recommendations.

4. Take the meter reading (room surface reflectance). Reflectance is the ability of the room's surfaces, such as its walls, ceiling and floor, to reflect light. Knowing reflectances will help you by revealing how much room surface dirt and surface finishes are absorbing light instead of reflecting it. To measure reflectance, get a footcandle reading holding the meter flat against the wall, with the sensor facing out into the work area, and note the reading. Then hold the meter about one foot (or one meter if measuring in metric lux) from the wall, with the sensor facing the wall, and take another reading. Divide the second reading by the first reading to obtain a reflectance value for the surface:

Second Reading ÷ First Reading = Reflectance (%)

To test the accuracy of your reading, you can use a white test card that has a known reflectance of 90 percent. Hold the test card against the wall and perform your test. If your results aren't near 90 percent, the meter needs to be calibrated.

5. If outdoor light cannot be prevented from entering the room, take one reading with the lights on, then turn the lights OFF and quickly take another reading. The difference between the readings (reading with lights on minus reading without) is the light level produced only by the lighting system.

Be sure to measure light levels at a variety of workplace locations, including specific task locations and randomly throughout the room, between luminaires and in corners. Utilize the IES procedure for conducting average illuminance measurements. Be sure to record the locations of readings for the baseline case so the procedure can be repeated after the wash and relamp. Adhesive labels can be used to mark measurement locations and corresponding values.

Chapter 2

An Introduction To Light

If electric power was all that mattered, we would all be
working in the dark.

—Leslie M. North, lighting consultant

WHAT IS LIGHT

Light is radiant energy that travels in waves composed of vibrating electric and magnetic fields. Light waves have both frequency and length; the ranges of frequency and wavelength differentiate light from other forms of radiant energy such as heat and radio waves. These properties are expressed relative to the electromagnetic spectrum.

Certain light waves comprise a portion of the spectrum called the visible light spectrum. Visible light is capable of exciting the eye's retina, producing a visual sensation that we call vision. The process of seeing, therefore, requires at least one functioning eye and visible light.

The eye, in fact, does not see objects; it sees only light that is emitted from a source, transmitted through a material, or reflected off of a surface, allowing us to perceive a source, a material, a surface. The object is visible because of contrasts of light. Its colors are perceived because those colors are contained in the light.

It is important to understand light because if we understand it, we can control it. The lighting industry has developed a series of metrics enabling practitioners to measure, understand and predict the behavior of light in a given application. In design, these metrics become tools, enabling the designer to control the behavior of light.

The most fundamental are light output (lumens), light level (footcandles or lux) and luminance (candela). Many retrofits focus on light output and light level, although luminance is actually the most important of these three basic elements because this is the light that we can actually see.

Light output, also called *luminous flux*, is the quantity of light exiting a light source or luminaire, measured in *lumens*. *Light level*, also called *illuminance,* is the resulting quantity of light falling on the tasks in the space, measured in *footcandles,* or lumens per square foot (or *lux,* lumens per square meter, which can be calculated by multiplying footcandles by 10.76), after all light loss factors are considered. *Luminance,* sometimes confused with perceived brightness, is the resulting quantity of light intensity radiated from the task in a given direction, measured in *candela* per square meter, based on the reflectance of the task and the size of the surface from which the light is being emitted (formally the *projected area*).

GOOD TASK VISIBILITY AND PERFORMANCE

One of the most important functions of a lighting system is to support visual performance by ensuring appropriate visibility of tasks and objects the designer wants seen. The human eye can adjust to a wide range of light levels, including about 10,000 footcandles on a sunny day to about 0.01 footcandles under full moonlight. As Americans spend some 85 percent of their time indoors, it is essential that interior lighting provide adequate light levels for people to perform tasks with desired speed, accuracy and safety.

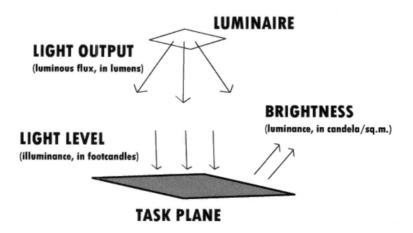

Figure 2-1. Simple diagram depicting relationship between light output, light level and luminance.

Formula for Task Visibility

The visibility of a given task is determined by the contrast of its brightness (luminance) or color against that of its surround. Other important factors include the size of the task, movement, time available to complete the task, and age of the person completing the task.

In short, the stronger the color contrast of a task against its surround, the easier it is to see. The larger the task is, the easier it is to see. Tasks that are motionless are easier to see than tasks that are moving, especially if that movement is fast or unpredictable (although movement can aid in detection, such as a person raising their hand during a seminar). If the user has a longer time to complete a visual task, less light may be needed on the task.

These factors are interdependent, so if one factor is strengthened, it can make up for weakness in another. They are also typically out of the lighting designer's control, however. But there is an important factor the designer does control, and that is luminance.

The Critical Role of Luminance

Again, luminance is light intensity emitted in a given direction after light exits a light source and falls on a task. The eye perceives light as brightness, the subjective experience of the light the eye sees. The higher a task's luminance relative to its surround, the more visible it will be. Note that the eye takes time to adjust to luminance changes in a process called adaptation, during which vision may be impaired to an extent, which may be a factor in some applications such as building entrances. The eye typically adapts to higher luminances quickly and lower luminances slowly.

Luminance is determined by the interaction of light level and task reflectance. Luminance contrast is determined by the luminances of the task and the source/task/eye geometry.

Generally speaking, increasing light levels falling on a task will increase luminance, which will increase visibility. As light levels are a function of lumen output, the lighting designer can, to an extent, influence task visibility by controlling light levels. Different tasks call for different *vertical* and *horizontal* light levels based on different combinations of size, movement, time, etc.; the Illuminating Engineering Society (IES) has developed light level recommendations for a wide range of tasks. For a description of the light level determination procedure and recommended values, consult the latest edition of the *IES Handbook* (at the time of writing, a new edition was tentatively scheduled for publication at the beginning of 2011).

Task and adjacent surface *reflectance*, expressed as a percentage of light reflected versus absorbed, is also important. The reflectance of the task and its surround should enable resulting luminance to be within recommended luminance ratios. Walls, ceilings and objects in the space, meanwhile, should have a high reflectance to produce desirable interreflections and improve average light levels and visual comfort. Reflectance should not be confused with glossiness (specularity); glossy surfaces may appear brighter to the eye, but that brightness is more likely to produce unwanted glare. For walls and ceilings, light-color matte finishes are often desirable.

Source/task/eye geometry—or the angular relationships between the light source, task and the user—is a determinant of luminance contrast. This is important to understand because sometimes lighting problems are not always problems with the lighting system. In some cases, users may be located or oriented in such a way they are more exposed to glare; the problem may be solved upfront by smart layout of task locations, relocating or reorienting the task, or moving the source if the luminaire is portable.

Other Considerations

When ensuring good visibility, lighting designers must take care to avoid visual fatigue, glare and shadows. Visual fatigue and glare are covered in the next section, where we will be talking about issues related to visual comfort. *Shadows* can be useful for providing spatial depth and defining details but unwanted shadows can product two negative impacts on task visibility. First, body shadows can reduce task light levels by obstructing light intended for the task. Second, the shadow may cover and therefore visually obscure the task itself. Point sources are more likely to produce shadows. Linear lighting and localized supplementary task lighting can help reduce shadows.

A final factor that is important to ask visibility is the age of the user. As people age, the pupil grows smaller for a fixed ambient light level. The eye exhibits a progressively lesser ability to focus on near objects, a condition called presbyopia. An average 50-year-old may need double the light level of an average 20-year-old to transmit the same amount of light onto the retina. What's more, older eyes are more susceptible to experiencing disability glare and take longer to adjust to changes in light levels (such as from one room to the next).

As can be seen, the luminous environment is much richer and more complex than simply delivering enough lumens to a horizontal workplane. In many retrofit situations, a considerable amount of trust is given to the original system designer, as existing light levels are either maintained or, in the case of older systems, reduced to current standards. In a relighting or new construction project, the lighting designer will review the application with a fresh eye and optimize task visibility without a requirement for allegiance to the existing design.

In the Spotlight: The Inverse Square Law

Light level describes the amount of light from a light source falling on a given surface area. As the light source and the surface move farther apart, the amount of area covered by the light source increases. As lumens remain constant but the area increases, the amount of light falling on each portion of the area decreases according to the Inverse Square Law:

$$E = I \div D^2$$

Where:

E	= light level (footcandles)
I	= intensity in candelas towards point P
D	= distance in feet from the source to the surface

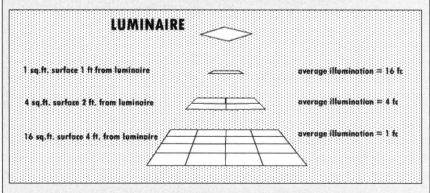

Figure 2-2. Example of Inverse Square Law.

The formula applies to instances where the receiving surface is perpendicular to the source direction. If light is incident at some other angle, the formula changes:

$$E = (I \cos \theta) \div D^2$$

Where:

E   =   light level (footcandles)

I   =   intensity in candelas towards point P

D   =   distance in feet from the source to the surface

θ   =   angle of incidence

Light intensity, measured in candlepower along a given solid angle, remains constant because while the area under the source increases, the solid angle is unchanged with distance.

MAXIMIZING VISUAL COMFORT

Lighting quality goes beyond task performance to address other needs related to aesthetic judgment, mood and atmosphere and visual comfort, all of which contribute to user satisfaction and task performance.

Aesthetic judgment and mood and atmosphere, often left out of the retrofit process, become primary considerations in a relighting or new construction project, covering patterns of light, room surface brightness and uniformity. These topics are covered in detail in **Chapter 3**.

Visual comfort generally means avoiding excessive luminance or contrasts in luminance that can cause either visual fatigue or glare.

Tired Eyes and Uniformity

Visual fatigue is the result of long exposure to excessive non-uniform contrast in a space, particularly when performing critical tasks, caused by *adaptation*, or the eye continually adapting to one level of luminance to another. For this reason, relatively uniform light levels are recommended for office, classroom, industrial and similar environments, where users perform sustained and/or critical visual tasks. Harsh patterns should be avoided. Another lighting problem that can contribute to visual fatigue is flicker.

Uniformity entails a relatively consistent distribution of illuminance and light intensity across the task plane, an imaginary plane placed on the level of primary tasks in the space. In an open office, this may be the desktop. In a hallway, the task plane may be the floor. There should also not be excessive contrasts between horizontal tasks and vertical surfaces in the field of view, so as to reduce stress on the eye due to adaptation.

Emotionally, the presence of good uniformity of light intensity in the visual environment, combined with higher light levels and a lower intensity at the room perimeter, can promote a public "work" atmosphere. Uniform distribution with high light levels and bright walls and ceilings can enhance the sense of spaciousness.

Task luminance should be higher than its surround to focus attention but not so high that uniformity is lost and visual fatigue may result. As a guide, IES provides recommended luminance ratios for various applications. Generally, up to three times more light on a workplane task than its immediate surround will direct attention to the task but not cause visual fatigue.

The Types of Glare

Excessive luminances or luminance contrasts may also lead to a visual sensation called *glare*, which, even though it is also light, can impair or disable vision rather than enable it. Glare is categorized as several different types according to its effects.

Direct glare is caused by directly viewing a light source, such as a bright window or an unshielded high-brightness lamp. *Reflected glare* is caused by light reflected from a surface, such as a veiling reflection on a glossy magazine or computer screen.

Disability glare is particularly problematic because vision is virtually disabled. *Discomfort glare* occurs when glare sources in the field of view produce a sensation of irritation in the eye. One type of discomfort glare that may occur in open office plans, classrooms and similar environments is *overhead glare*, which, unlike the usual discomfort glare, does not occur within the field of view. Instead, it is usually experienced by people performing heads-up tasks, such as office workers typing on keyboards or students observing a teacher, under recessed 2x4 luminaires. The sensation of brightness does not abruptly cut off but instead gradually fades below the line of sight to the glare source, producing a sensation of glare due to light scattering at the eyebrow and cornea of the eye. Light may also be reflected into the eyes from the face around the eyes, such as the nose and cheekbones.

If glare is present, consider the source, the task and their relationship with each other and the user. Changing one or a combination of these should be able to solve most glare problems. **Table 2-1** provides a summary of IES recommendations for mitigating glare problems in office, education and industrial spaces.

Regarding the source, avoid direct sunlight penetration into the space. Bright windows can be mitigated by placing light on adjacent walls to reduce contrast or by providing shading. Avoid excessively bright lamps in luminaires where the lamps are visible. If bare lamps will be a potential source of glare, consider low-brightness luminaires.

Regarding the task, consider placing light on the task's surround to reduce contrast. Consider changes to the source/task/eye geometry that may solve the problem, such as relocating or reorienting the task. Reflected glare on computer screens may be less of a problem than just a few years ago as computer screens have advanced—larger radius of curvature (i.e., less curvature), improved screen brightness, anti-reflectance technology, flat screens, positive contrast in software, etc.—thereby reducing the risk of unwanted reflections.

Note that while glare is typically best avoided, points of high brightness can be welcome in some cases, such as tiny points of light, or "sparkle," used to convey an atmosphere of elegance. Examples include bright highlights on silverware in a fine restaurant or in a chandelier mounted over a grand lobby.

Figure 2-3. If the bare lamp is visible to the user's eye, even if it is visible above the natural cutoff angle of the eyebrow, users may find it comfortable to shield their eyes with their hand or a baseball cap to avoid their exposure to overhead glare. Graphic courtesy of Naomi Miller.

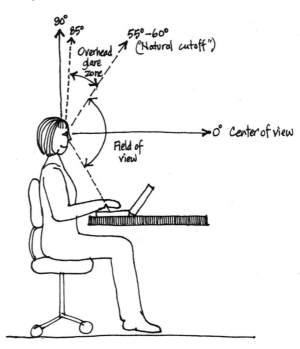

Table 2-1. IES recommendations for mitigating glare.

	IES RP-1-04, Recommended Practice for Office Lighting	IES RP-3-00, Recommended Practice for Lighting Education Facilities	IES RP-7-01, Recommended Practice for Lighting Industrial Facilities
General glare (excessive luminances or luminance contrasts in field of view)		1. Avoid high-brightness contrasts in field of view 2. Increase brightness of background to very bright objects 3. Raise the glare source	
Direct glare (caused by high luminances or unshielded light sources)	1. Reduce relative brightness of luminaires and windows 2. Shield view of bright sources 3. Increase brightness of background to very bright objects	1. Low-brightness luminaires 2. Control window luminances (e.g., shades, louvers)	1. Reduce the luminance or area of the glare source (e.g., luminaire shielding) 2. Raise the glare source further above line of sight 3. Increase ambient illuminance
Reflected glare (caused by high luminance sources or veiling reflections)	1. Locate luminaires or shield windows to avoid problems with glossy surfaces 2. Large area, low-luminance luminaires or indirect lighting 3. Ant-reflection coatings or filters for computer screens	1. Place illuminance on both sides of task 2. Special luminaire optical designs 3. For VDT applications, use monitor with diffuse reflecting screen and dark test on bright background	On VDTs: 1. Total cutoff of light source images 2. Change VDT orientation and position 3. Improved-contrast screens 4. Block view of light sources in offending zone
Disability glare (veiling luminance covers retina)			1. Reduce illuminance on eyes 2. Raise glare source
Discomfort glare (luminous image has much higher luminance than balance in field of view)			1. Size, luminance and source/eye/task geometry are factors

RENDERING COLORS AND SKIN TONES

Light is actually made up of wavelengths that the human eye interprets as colors. For an object to be perceived a certain color, that color must be present both in the light striking an object and the object reflecting the light towards the observer's eye. Because of this, choice of lamp can influence how colors are perceived in a space. This means that perception of objects in the space is, to an extent, controllable. In some applications, color is not very important, while in others it can be the most important lighting characteristic. For example, a recent survey of lighting designers and specifiers by the National Lighting Product Information Program showed that for retail applications, light source color properties are considered more important than any other light source criterion, including energy efficiency.

Color Tone

The *color temperature* of a light source, expressed in Kelvins (K), indicates the color tone of the light source itself and the light it emits. Light sources are generally classified as "cool" (>4000K), which appear bluish-white; "neutral" (3000K-4000K), which appear white; or "warm" (<3000K), which appear orangish-white (see **Table 2-2**). Warm light sources are more heavily laden with red and orange wavelengths, bringing out some flesh tones and richer content in objects that have

Table 2-2. Color temperature scale with approximate examples.

10000+K	North light (blue sky)
7000K	Overcast daylight
6500K	Daylight fluorescent lamp
6000K	Cloudy sky
5500K	Direct mid-summer sunlight
5000K	Direct sunlight, noon daylight
4500K	Sunlight in early morning or late afternoon
4000K	Clear metal halide lamp, cool white fluorescent
3500K	Neutral white fluorescent
3000K	Incandescent lamp, halogen lamp, warm white fluorescent
2500K	Sun at sunrise or sunset
2000K	Candle flame, high-pressure sodium lamp
1500K	Match flame

In the spotlight: Lighting influences taste of wine

Lighting influences how wine tastes and how much consumers are willing to pay for it, based on a series of three German experiments in which more than 500 people tasted white Riesling wines. The study report, "Ambient Lighting Modifies the Flavor of Wine," was published in the December 2009 issue of *Journal of Sensory Studies*.

The only significant variable in the experiments was the ambient lighting in the spaces where study participants sat. Researchers used a series of fluorescent lamps that produced red, blue, green or white light. People rated the wine's quality higher, in general, when they drank it in a room whose ambient lighting was red or blue versus green or white. They also found the test wine much sweeter and fruitier when sampled in a room illuminated by fluorescent lamps with a warmer color tone, and were willing to spend more for it.

---

warmer colors. Cool light sources are more heavily laden with blue and green wavelengths, enriching the visible color content of blue and green objects.

Color rendering

Color temperature describes the "whiteness," "bluishness", etc. of a light source—its warmth or coolness. However, it does not define how natural the color of objects will appear when lighted by the source. Two light sources can have the same color temperature, but render colors differently. The color rendering index (CRI), a rating scale with a maximum of 100, offers a metric to address this. For most common color temperatures, CRI uses a blackbody radiator as the reference for color rendering (assumed 100 CRI). Daylight has a CRI of 100, with incandescents coming in a close second. Incandescent lamps, however, are not a perfect light source for color rendering (they are weak in blue), so the CRI system has its drawbacks. It is, however, the only internationally agreed upon system for expressing a lamp's color rendering ability. It should only be used as an indicator of relative, not absolute, color rendering ability.

Specifying Color Quality

Specifying color quality appropriate for the application is essential. In a high-end retail application, good color quality can make products, especially clothing, appear more vibrant, truer and ultimately more

Table 2-3. Performance characteristics of major lamp types. Source: Craig A. Bernecker, PhD, FIES, LC.

Type	Correlated Color Temperature (K)	Color Rendering Index (CRI)
Incandescent	2800	90-95
Halogen	3000-3150	90-100
Fluorescent	3000-6000 (linear); 2700 (CFL)	50-90
Mercury Vapor	3000-6000	20-50
Metal Halide	3000-4500	60-70 (standard); 85-90 (ceramic)
High Pressure Sodium	2000-3000 (generally about 2500)	20-30

appealing. In an office application, color quality can facilitate social interaction by rendering faces more naturally and thereby helping to create a more productive and appealing work environment. In a hospital application, good color choices can make guests feel more welcome by imitating the warm lighting conditions of home, while rendering skin tones more effective to help detect illness (jaundice, rash, infection, fever, etc.). The list goes on.

Think about the application: Do you want it to be visually warm or visually cool? The right color temperature is often a matter of psychology—both preference and tradition—with due consideration for emphasizing (and not distorting) dominant colors in the space and properly rendering color contrasts that make tasks more visible or call out safety markings and instructions.

In cooler northern climates, people tend to prefer warmer light sources, while in warmer southern climates, people tend to prefer cooler light sources. At home, people tend to prefer warmer light sources, while at the office, cool or neutral light sources are preferred. Research also suggests that people are more accepting of warm color temperatures at lower light levels and cool color temperatures at higher light levels. "This makes sense when one remembers that traditional sources of low-level illumination for people were open fires, torches, candles and oil and gas lanterns—all sources rich in the red region of the spectrum," Gary Gordon notes in an article published in Architectural Lighting Magazine. "Traditional sources of high-intensity illumination have been the sun and sky, both relatively 'cool' colors of 'white' light."

Note that people need to look as good as objects, spaces and furnishings in the space; one study indicates that people (regardless of race) have a preference for neutral color temperatures (3000K-4100K) as rendering skin tones most pleasantly (Quellman and Boyce, 2002).

In many applications, the higher the CRI the better, with 80-100 being optimal for rendering colors more "naturally"—that is, how most people would expect them to appear, and >90 for color-important applications, such as retail applications where merchandise appearance and appraisal is critical. Many energy-efficient light sources, from T8 linear fluorescent to ceramic metal halide, offer good color quality. Note again that color rendering is also dependent on color temperature, so looking for the right combination for a specific project need is essential.

A Few Words of Caution on Color

For those specifying color characteristics of light sources, a few words of caution: While color temperature and CRI are important, neither fully describe the color characteristics of a light source. To more fully understand the color characteristics of a light source, consult the source's spectral power distribution curve, available from the manufacturer, which can be an important reference. That being said, however, there is no substitute for seeing the lamps in a given application. A separate mockup with each proposed lamp type is encouraged for color-critical applications because color quality is altered by room colors, luminaire characteristics, whether daylight is present, light levels and surface reflectances. Further, different manufacturers may rate their products very similarly (a few points difference in CRI may not be detectable), but when these lamps are seen side by side, there may nonetheless be slight visual differences, which is why manufacturers advise against mixing lamps from different manufacturers in the same application.

If task lighting is added to the design, make sure the lamp color matches the general lighting. If the project is a retrofit, the specifier may neglect to specify color characteristics, which can result in poor quality, so even for projects that are driven by energy, lamp color quality should be given due priority and specified. Note the impacts of daylight (does not mix well with warm light sources), color transmission quality of the window glass (can distort color quality in the space), and color shift in the source during normal aging and dimming (such as white-light LED luminaires, which shift in color as they age). Finally, the proliferation of color temperatures available in fluorescent lighting has made it easy to make mistakes such as mixing warm and cool lamps in the same application, which is why a maintenance plan, which includes a lamp schedule, can be an important tool to help maintain design integrity over time.

Besides source color, consider the colors used in the room surface finishes, furnishings, partitions, etc. in the space. Lighter (non-glossy) colors are more reflective (but not glaring), which can raise average light levels and reduce demand on the lighting system. For this reason, lighter colors are recommended for large area surfaces in many typical applications such as offices and schools, while darker finishes are recommended to be limited to accents.

Chapter 3

Lighting Design

Design is all about human beings.
—Howard Brandston

DESIGN MATTERS

As a new construction or relighting project is an opportunity to design or redesign the lighting system according to modern best practices, it is important to understand lighting design. Lighting design is the process of developing lighting that enables the safe, productive and enjoyable use of buildings and spaces. The complete process includes programming, schematic design, design development, contract documentation, bidding and negotiation, construction and post-occupancy evaluation. Fundamentally, it involves identifying the lighting goals and selecting the right combination of equipment and techniques that will satisfy these goals. This combination can vary from project to project, making lighting design a creative process that is often as much art as science.

Several years ago, this book's author had the honor of working with legendary lighting designer Howard Brandston to edit his book *Learning to See: A Matter of Light*, which was subsequently published by the Illuminating Engineering Society in 2008. At 138 pages, it is a slim book but ambitious in its comprehensiveness, covering the science of light and the art of lighting. It is Brandston's gift to the design community, a playful but incisive distillation of more than 50 years of experience designing and teaching lighting: a career boiled down into insight and advice.

If the book, in turn, might be boiled down to a single statement, it might be Brandston's trademark question, "What is it you wish to see?" An iconoclast at heart, he challenges today's lighting designers to transcend the rules and conventions to viscerally connect with the project and its users. Put yourself in the role of the customer, he says.

The worker. The traveler. You are here, going there. What is it you wish to see? In this way, he reminds us of the artistic nature of lighting and to think about light as an artist, how to use it to achieve the desired emotional response among the people who will use the light.

The "Who" of Lighting

It all begins with the "who" of lighting: Who are the ultimate users of the lighting, how will they use the space, and what are their lighting needs? Programming also involves understanding the task and space characteristics, noting any important features, materials, finishes, architecture and energy, maintenance and safety issues. In an existing building, this information can be produced during an audit that includes surveying user preferences, not just dollars and watts, and a walkthrough that includes an analysis of lighting quality factors, not just numbers of lamps, ballasts and luminaires. All parties involved should share needs and design ideas in practical terms that are easily understood by everyone: The design process should be creative, but the design intent should not be open to interpretation.

The Emotional Environment

Begin with the most appropriate emotional environment for how the space is used: Should the lighting make the space appear public or intimate? Relaxed or formal? Spacious or small?

Humans respond to light physiologically and psychologically. It starts with the fundamental premise that brightness focuses attention. The human eye is naturally drawn to areas of brightness and brightness contrasts in the field of view. Contrast is even more important than brightness, which is why lighting designers consider darkness a very important tool in designing with light.

This premise is rooted in our scientific understanding of phototropism, the tendency of humans, animals and plants to seek light. People will reflexively orient themselves to squarely face sources of high brightness and strong brightness contrasts as long as they are not uncomfortably glaring.

In the 1970s, Dr. John Flynn conducted experiments on people's subjective responses to different lighting conditions. In one study, participants entered a cafeteria and showed a preference to face the entrance. After wall lighting was added, participants entering the cafeteria changed their orientation and preferred to face the bright walls.

In another study by Taylor and Sucov in 1974, study participants entered a room through a curtained entrance and read instructions on a room divider telling them to go to the other side and complete a task as part of a study of consumer product lighting. Unaware of the true intent of the study, they entered the room by choosing to go either left or right to get past the room divider. When both paths were lighted equally, seven out of 10 participants went to the right. When the path to the left received a higher light level, three out of four went to the left. Light not only focuses attention, it can help with wayfinding.

And not only do people respond to light behaviorally, they also respond psychologically to patterns of light in a space.

Dr. Flynn developed a series of criteria that can be used to evaluate how people will respond to different lighting arrangements based on a series of criteria: overhead (horizontal) versus perimeter (vertical) emphasis, uniform versus non-uniform distribution, bright versus dim light levels, and visually warm versus cool color tones (see **Figure 3-1** for an illustration of these effects). In a series of simple experiments, people were exposed to different lighted spaces and asked to rate their subjective impressions on a scale between "pleasant" versus "unpleasant," "spacious" versus "confined," "relaxed" versus "tense," and "visually clear" versus "hazy."

Intense direct light from above, coupled with non-uniform light distribution, can create an impression of tension suitable for a high-end hotel lobby. Lower overhead lighting with some lighting at the room perimeter, coupled with non-uniform distribution, can foster a relaxed atmosphere (see **Table 3-1**). Bright light with less intense perimeter lighting, coupled with uniform distribution, can promote a public "work" atmosphere suitable for an office. Bright light with lighting on walls and possibly the ceiling, coupled with uniform distribution, can create a sense of spaciousness, while low light levels at the activity space with a little perimeter lighting and dark areas in the rest of the space, coupled with non-uniform distribution, can create an intimate atmosphere suitable for a high-end restaurant.

Placing the Light

Consider what to light in the space: What surfaces should be lighted? Should the lighting emphasize people and tasks, the architecture or both? What focal points require emphasis in the visual hierarchy for either visual interest or wayfinding?

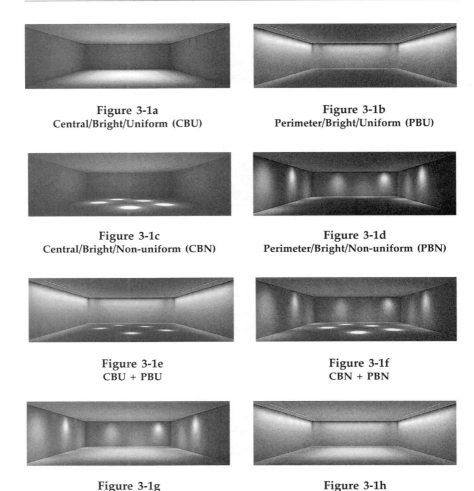

Figure 3-1a
Central/Bright/Uniform (CBU)

Figure 3-1b
Perimeter/Bright/Uniform (PBU)

Figure 3-1c
Central/Bright/Non-uniform (CBN)

Figure 3-1d
Perimeter/Bright/Non-uniform (PBN)

Figure 3-1e
CBU + PBU

Figure 3-1f
CBN + PBN

Figure 3-1g
CBU + PBN

Figure 3-1h
CBN + PBU

Figure 3-1. Dr. John Flynn's research indicated that people form subjective impressions of built environments independently of the architectural context, categorizing them in terms of overall central or perimeter emphasis of the lighting. Shown here are eight renderings of a room, developed based on Flynn's studies, with central, perimeter and combined lighting, bright illumination as a constant (compared to dim), and both uniform and non-uniform light patterns on surfaces. One can see at a glance how different lighting configurations produce different appearances of the space with different resulting subjective impressions. Images courtesy of Dr. Robert Davis.

Table 3-1. A space's lighting defines its personality and how people perceive it, which in turn affects how they feel about being there. Below are various lighting effects that can take the same space and transform it into different environments.

Psychological impact	Lighting effect	Light distribution
TENSE	Intense direct light from above	Non-uniform
RELAXED	Lower overhead lighting with some lighting at room perimeter, warm color tones	Non-uniform
WORK/VISUAL CLARITY	Bright light on workplane with less light at the perimeter, wall lighting, cooler color tones	Uniform
SPACIOUSNESS	Bright light with lighting on walls and possibly ceiling	Uniform
PRIVACY/INTIMACY	Low light level at activity space with a little perimeter lighting and dark areas in rest of space	Non-uniform

As stated above, the principle of phototropism indicates that the eye is drawn to the brightest point in the field of view. Light, therefore, provides visual cues to draw the attention of occupants to objects or surfaces through the contrast of brightness (and/or color) between the object/surface and their surround as well as other objects/surfaces in the space. Overall, if light is concentrated on objects and architecture, then people will focus their attention on them. If light is uniformly concentrated on horizontal surfaces, such as desktops in an office, then people and activities become the focus of attention. In some cases, the luminaire itself may be required to become a focal point, a decorative luminous element in the space.

The designer therefore must determine the space's focal points to decide what objects and surfaces need to be lighted. Strong contrasts are visually stimulating and can create points of interest within a sharper visual hierarchy, although too high a contrast will produce glare and should be avoided. As an example, consider statues in an art gallery, where strong contrasts highlight the artwork and are visually exciting. In a space where people need to perform demanding visual tasks, however, such as an office or classroom, such contrasts can be visually fatiguing, and so the focal point is the work surface, which is lighted uniformly, although a little variation may be carefully provided (such as daylight or perimeter lighting) to relieve the visual monotone.

Many applications have common needs regarding the focus of light in the space. In a retail environment, key merchandise on display represents focal points that should be given lighting emphasis so that shoppers focus on it and find it more appealing. In an office or classroom, horizontal spaces where work is performed are the focus, with some light on walls and ceilings to create a sense of spaciousness. In a grand lobby of a large office building, the architecture becomes the focus along with light that provides visual cues to the security desk, elevators, exit and other points of interest such as logos or artwork. In

Figure 3-2. Lighting can affect how spaces are perceived. To study the effect of various light distributions in a typical office space, lighting designers Leslie North, PE, LC, LEED-AP and Carla Bukalski, PE, LC created a sample space and lighted it to 50 footcandles four different ways using lighting visualization software. Here we see a private office lighted using lensed troffers (a), downlights (b), parabolic troffers (c) and linear indirect (d). Each provides the same light level but creates different visual environments. Which space appears most vi-

sually comfortable to you? Which appears brightest and most spacious? In which space would you most like to work? Images courtesy of Leslie North.

an upscale restaurant, the focus of the light could be the small space between two people having dinner so as to create intimacy. And in a nightclub, the focus could be kinetic, constantly shifting to create tension and stimulation.

The Case for Wall Lighting

In many applications that are typical retrofit candidates—offices, classrooms, etc.—the lighting needs are fairly simple. In terms of lighting quality, the lighting system should provide bright and uniform illumination appropriate for the space's visual tasks, render colors and skin tones naturally, and create a visually comfortable environment without direct or reflected glare.

These spaces are particularly vulnerable to poor lighting quality, however, partly because of the green design movement. Some of the challenge stems from traditional lighting design's focus on providing

In the Spotlight: Common formulas

$$\text{Average Maintained Illumination (Footcandles)} = \frac{\text{Lamps/Luminaire} \times \text{Lumens/Lamp} \times \text{No. of Luminaires} \times \text{Coefficient of Utilization} \times \text{Light Loss Factors}}{\text{Area in Square Feet}}$$

$$\text{Average Maintained Illumination (Footcandles)} = \frac{\text{Total Lamps} \times \text{Lumens/Lamp} \times \text{Coefficient of Utilization} \times \text{Light Loss Factors}}{\text{Area in Square Feet}}$$

$$\text{Average Maintained Illumination (Footcandles)} = \frac{\text{No Lamps in 1 Luminaire} \times \text{Lumens/Lamp} \times \text{Coefficient of Utilization} \times \text{Light Loss Factors}}{\text{Area in Square Feet Per Luminaire}}$$

$$\text{Required Lumens/Luminaire} = \frac{\text{Maintained Illumination} \times \text{Area (Square Feet)}}{\text{No. Luminaires} \times \text{Ballast Factor} \times \text{Coefficient of Utilization} \times \text{Light Loss Factors}}$$

Light Loss Factors = Ballast Factor x Lamp Lumen Depreciation Factor x Luminaire Dirt Depreciation Factor x Lamp Burnouts Factor

And if determinable... x Luminaire Surface Depreciation Factor x Room Surface Dirt Depreciation Factor x Luminaire Ambient Temperature Factor x Supply Voltage Variation Factor x Lamp Position Factor (HID systems) x Optical Factor, etc.

sufficient light levels on the horizontal task plane. In a corridor, this might be the floor. In an office or classroom, the tops of the desks. In trying to squeeze every last watt out of the lighting system, these light levels can be pushed as low as possible with a system that only provides direct light onto the workplane; the result may be high energy efficiency but a gloomy atmosphere that does not serve architecture or people.

A solution might be found in wall lighting provided by general lighting with an indirect distribution component, "volumetric" distribution recessed general lighting (which distributes a higher intensity of light at higher angles, illuminating walls), placing direct lighting close to walls, and/or dedicated wall lighting such as wall washers and wall/slot systems. Placing light on walls, considered best practice in contemporary lighting design, can produce two significant benefits. First, it defines the spatial form, making the space appear brighter and more spacious, while focusing attention on people and architecture (see **Figure 3-3**). Second, when combined with high-reflectance surfaces, it produces useful interreflections that can enhance light levels and visual comfort.

Typically, vertical surfaces are an occupant's first impression of a space. Based on Dr. John Flynn's studies, emphasizing lighting on vertical surfaces produces predictable and consistent favorable perceptions of the space. Flynn found that brightness on walls tends to produce impressions of spaces as being "pleasant" (see **Figure 3-4**). He also found that wall lighting can contribute to impressions of a space being "public," "spacious," "tense" ("open for business," or "businesslike") and "visually clear," particularly when light distribution is uniform.

Besides directing attention and influencing aesthetic judgment of a space, wall lighting can also contribute to visual clarity and comfort. One of the most beneficial applications of wall lighting in applications with sustained, demanding visual tasks is to turn walls into area light sources, producing interreflections that can increase light levels, uniformity and visibility while reducing shadows and strong contrasts. Increasing uniformity reinforces impressions of spaciousness, alertness and visual clarity. Increasing distribution on vertical surfaces increases their visibility while improving facial recognition, which can aid face-to-face communication. Reducing shadows and strong contrasts, meanwhile, can increase visual comfort. A good example would be a windowed space where there may be an uncomfortable brightness contrast between the windows and adjacent wall spaces; placing light on these walls softens the contrast.

Figure 3-3. Wall lighting illuminates the spatial envelope providing the dominant boundaries of spaces, providing a pleasing luminous backdrop to people, objects and activities. Photo courtesy of Litecontrol.

Quantity of Light

A primary function of a lighting system is to enhance visual performance by increasing the visibility of objects and tasks: How much contrast and color difference is there between task details and their background? How much time does the user have to perform the task? Is the task moving—and if so, how fast and is the movement predictable?

While this stage of the design is of course critical, note how far we have come before determining target footcandles on work surfaces based on recommended formulas. In a typical lighting design, the designer references IES-recommended light levels, selects generic luminaires and lays out the luminaires in a ceiling grid. But as we have seen, best practices transcend simple calculation of light levels on the workplane.

The visibility of a given task is partly a function of luminance contrast (and/or color difference) between a task and its surround. In

Figures 3-4. This BEFORE (top) and AFTER (bottom) comparison emphasizes the qualitative importance of Room Surface Brightness in addition to Points of Interest, Modeling of Faces and Objects, Illuminance on the Task, and Appearance of Space and Luminaires. Note that finish selection also contributes to room surface brightness. Figure 3-4a courtesy of Community Protestant Church. Figure 3-4b courtesy of Leslie North/Aurora Lighting Design, Inc.

review, light level, measured in footcandles or lux, is the quantity of light falling on a task, while luminance, measured in candelas, is the intensity of light emitted from a surface, either directly from a luminaire or reflected back to an observer's eye. Luminance contrast (and/or color difference) is required for vision, with visibility increasing as contrast increases.

To increase contrast and thereby increase visibility for a given task, the designer can increase the quantity of light falling on it, increase its surface reflectance, or change the light source (specifically, specifying a lamp with a different spectral composition). Delivering more footcandles to the task typically improves visibility (depending on the directional reflectance characteristics of the task's surface material)—until excessive luminance contrast results in glare, which of course should be avoided. Assuming glare is avoided, there is also a point of diminishing returns where raising light levels will begin to make little difference in visual performance.

There are other factors that are often outside the designer's control but should be accounted for when determining light levels for a given task. One is source/task/eye geometry, which expresses the angular relationships between the user, the task and the luminaire; changing any of these relationships can affect how visible a task is. Then there is size, movement, time: Generally, higher luminance can improve visibility and/or detection if the task is visually small, is moving quickly and/or in an unpredictable direction, and/or must be detected and completed quickly and with a high degree of accuracy. Special safety hazards should also be taken into account. Finally, the user's age should be considered, as the pupil becomes progressively smaller as the eye ages; an average 50-year-old may, in fact, require twice the light levels falling on a task as a typical 20-year-old to gain the same amount of light on the retina.

When determining light levels, the lighting designer may need to consider horizontal and vertical light levels. Horizontal light level is the amount of light falling on a horizontal workplane (e.g., a desktop in an office), while vertical light level, of course, is the amount of light falling on a vertical surface (e.g., the titles on the spines of books placed on a bookshelf). Light levels should be determined in accordance with IES recommendations, which are based on requirements for a typical user in a common environment for a given task. The ASHRAE 90.1 energy standard, on which many commercial building energy codes is based, is designed based on an assumption that light levels for various space and building types will achieve IES recommendations for the given maximum allowable lighting power density, giving IES recommendations, in one sense, the force of law.

IES publishes recommended practice guides for various vertical markets, from education and healthcare facilities to office buildings and roadway lighting. These guides provide recommended procedures for determining minimum maintained light levels for various tasks and spaces that are typical for the given vertical market.

Concept, Then Design

The lighting concept and design follows from satisfying the above project needs. The lighting concept is an expression of the design intent: Write it down. If the owner wants the lighting to make an office space appear public, bright and spacious, then the basic design concept might be, "A direct/indirect lighting system will be used to provide task illumination for users while illuminating upper walls and the ceiling to make the space appear bright and visually larger." In a reception area attached to the office, another goal might be to attract visitor attention to the reception desk and a wall-mounted company logo behind it. In this case, the basic design concept might be, "Light intensities on the reception desk and company logo artwork behind it will be at least three times higher than the general ambient lighting, so as to attract occupant attention."

The resulting design may require different lighting techniques and subsequent combination, or layering, of these techniques in the space. The most basic layers are general/ambient, task and accent, but others can be used to create a variety of effects within a lighting composition.

LIGHTING TECHNIQUES

By understanding basic lighting techniques and how to layer them successfully, designers of lighting systems can help owners realize their lighting goals.

The selection of the right techniques depends on the requirements of the application. It may be desirable to provide strong accent lighting in a retail environment to accentuate and dramatize key merchandise,

In the Spotlight: The Squeeze on Design

Energy codes are undeniably becoming more restrictive with each generation of standards. By effectively dictating a narrowing set of choices, they are gradually evolving into design guides. Meanwhile, green building rating systems such as LEED require and encourage going beyond energy code. As mentioned in an earlier chapter, LEED is already required in public construction in more than 30 states and 130 cities. As the first green construction standards enter the national code stage, green construction practices may become code—that is, applied to construction and renovation projects with the force of law.

In the July/August 2009 issue of Illuminate Magazine, lighting industry consultant Kevin Willmorth writes that with all of the rules and regulations impacting today's lighting designs, "Lighting design is beginning to feel more like regulation engineering than art." The increasing complexity of lighting caused another lighting designer to quip that her job felt more like accounting than lighting design.

The conflict between energy efficiency and lighting quality is difficult to assess because lighting quality itself is more difficult to measure than efficiency. But remember: While energy savings are more easily measurable, user satisfaction can have a much greater impact on the bottom line. In 2006, I conducted a survey of electrical engineers, architects and lighting designers about energy-efficient lighting. When asked to rate various drivers and barriers to energy-efficient lighting, the respondents rated "provides the greatest value to the client" as a major demand driver, but also cited "unacceptable tradeoffs with lighting quality" as a major barrier to adoption. One can conclude from this that respondents place a high value on energy-efficient lighting, but are concerned about potential quality compromises.

Energy codes are considered a societal good. They contribute to more sustainable use of finite energy resources and reduce air emissions while increasing business competitiveness. Lighting quality is, arguably, another societal good. As energy codes grow more restrictive, they can reduce artistic freedom, reduce choice and possibly negatively impact lighting quality. These risks are accelerated in LEED projects where code is leapfrogged.

For example, energy codes are designed on the basis of achieving minimum recommended light levels but quality lighting involves much more than light levels. Numerous design issues must be addressed, such as glare, room surface brightness, color and many others. For example, some of the most efficient luminaires can produce minimum recommend light levels but are categorically described by designers as "glare bombs."

In green projects, Naomi J. Miller, FIES, FIALD, LC, warns in the January 2009 issue of Illuminate Magazine, there is a tendency to emphasize low power densities and new technologies at the expense of lighting quality. One example is a trend toward lighting offices with narrow 4-ft. recessed slots, sometimes mounted on darker ceilings resulting in excessive contrast (glare) and distracting images even in some flat LCD screens. Another, she points out, is T5HO lamps misapplied in direct luminaires with the bare lamp visible to the user's eye, which is likely to result in overhead glare. Another is flicker, an occasional problem with older magnetic fluorescent ballasts that was eliminated with electronic ballasts and has been resurrected with some LED product. Another is high Visible Light Transmittance (VLT) windows, which can deliver poor color quality (heavy in green wavelengths). And another is gloomy lighting.

"A few folks enjoy working in gloomy spaces, but they are a small minority," she writes. "Leave the atmospheric look to funky restaurants, theatres and hotels. LEED offices, schools, healthcare and industrial workspaces need to look cheerful, and illuminating walls and ceilings are the best tools we have to do it. You can do it with fewer watts if the wall and ceiling surfaces are lighter in color, whether it's wood or concrete or paint."

Willard L. Warren, PE, FIES, principal of Willard L. Warren Associates and a regular columnist for LD+A Magazine, suggested in his August 2009 column that goals like LEED certification should not trump lighting quality, which is needed for people, not points.

"When relighting an existing space, don't be superficial," he writes. "All the prizes and awards in lighting are based on beating the lighting power density (LPD) limits of the energy codes. This may have less value to the owner of a business who is more interested in improving the performance and reducing the errors of the employees. Winning prizes is great, but improving performance while lowering energy bills is priceless."

Energy codes, green building codes and programs such as LEED are beginning to exhaust much of what lamps and ballasts can do; the next level of energy savings is expected to come from a combination of innovative design, daylighting and lighting controls. As the pressure to achieve high energy savings drives lighting to become more and more complex, the practice of achieving good lighting will become more and more demanding, with fewer design choices.

Willmorth argues that lighting's growing complexity is a bad thing. But also surprisingly good. "As a designer who has always been frugal with energy, having regulations limiting customer options is actually liberating, as it pushes their choices toward more conservative and sustainable solutions without all the argument and selling," he writes. "For the time spent on compliance checks, there is now time saved not making presentations to convince a reluctant customer to 'do the right thing.' While the market feels more complex and controlled than ever, the opportunity to put the pieces together with emerging new technologies makes design as challenging and interesting as it has always been—constantly evolving and changing."

but in an office, such strong concentrations might prove visually fatiguing. Uplighting may work well in an intimate restaurant or to highlight bottles of alcohol in a bar, but may make people look sinister in the home or office. Sparkle and glitter may work well in a restaurant, but may prove distracting in an industrial space.

Most lighting designs feature fixed patterns of light, although irregular patterns, such as scallops, may be desirable for special effects. Other irregular contributions to illumination, such as daylight entering a room, can also be desirable. Light patterns from luminaires can be sharply defined or diffused. The beam spread should match the form of the object or surface being lighted. For example, a narrow, highly concentrated light beam from an accent luminaire may be suitable to light a mannequin in a clothing store, but not a wall or ceiling, which instead would warrant a wide, softer distribution.

Task Lighting

This is the layer that provides sufficient light levels for users to perform visual tasks. The lighting should be designed so that it provides sufficient maintained light levels for users to perform tasks at the desired degrees of speed, accuracy and safety, taking into account all light loss factors that erode light levels over time.

Task lighting may be part of the ambient illumination layer or localized lighting. Ambient lighting, also called general lighting, is usually provided by ceiling-mounted lighting equipment, and provides sufficient light primarily for the task of orientation (people walking around the space). As ambient lighting is usually diffuse and uniform, it can be bland; for visual interest, accent or other lighting techniques can be used to create focal points or other interest. Task lighting is usually provided by ceiling-mounted equipment, localized equipment such as portable or hardwired task lights, or a combination of the two, and provides sufficient light levels for performance of specific visual tasks.

Accent

Accent lighting is used to highlight key objects, displays, artwork, standout architectural features and special areas by making them significantly brighter than their surround (see **Figure 3-5**). As the eye is attracted to points of brightness in the field of view, focal points and a visual hierarchy can be established. Highlighting areas, meanwhile, can be useful for wayfinding, as in our earlier example of the reception desk. Color contrast can also be effective (see **Figure 3-6**).

When an object is to be accented, we typically want to highlight the object only and avoid spill light on the surround, which would reduce the contrast we need, so focused, directional luminaires are often used to produce effective accent lighting, particularly when smaller objects are involved. Various light sources and luminaires offer a range of distributions from wide to pin spot, enabling a tight focus on even very small objects; examples of light sources include halogen (low or line voltage), ceramic metal halide, compact fluorescent and LED. To make the object very bright, we will need sufficient "punch"—a high level of intensity in the luminaire's center beam. Intense directional lighting will create shadows and resulting strong texture, which can be emphasized using point-source lamps.

Accent illumination should make sense—i.e., be used to highlight special objects and features. For example, if a statue in a corner is moved but the accent light is still operating, then we are drawing attention to an empty corner, which is visually confusing and distracting. For this reason, if the illuminated object is likely to be moved, it would be useful to have luminaires that can be either re-aimed or moved, such as track lighting or aimable downlights.

One interesting type of accent lighting is called framing, involving a recessed or surface-mounted framing projector with adjustable shutters, which provides precise focus on, for example, a painting on a wall. The effect is highly dramatic as the painting becomes very bright and its surround dark, resulting in high contrast and visual interest.

Downlighting

Downlighting places light on objects or surfaces below a luminaire that aims light downward (see **Figure 3-7**). In some applications, downlights can be used to make a space appear smaller and more intimate. Intense, non-diffuse downlighting can be used to create an exciting atmosphere by producing high contrasts. It should be avoided, however, in spaces with critical prolonged tasks, as high contrasts are exciting but can be visually fatiguing over time. It should also be avoided in spaces where social communication is critical, as the shadowing produced by non-diffuse downlighting can render faces harshly.

Downlights do not have to be placed in a uniform arrangement, but the arrangement should be organized and "make sense." Higher mounting heights may require narrower beam spreads to avoid glare.

Figure 3-5. Accent lighting. Photo courtesy of W2 Architectural Lighting.

Figure 3-6. These luminous color accents call attention to these bookshelves. Photo courtesy of Philips Color Kinetics.

Downlights placed close to a wall can produce tall and thin scalloping, which should generally be avoided unless desired for a particular aesthetic reason. Typical light sources include compact fluorescent, ceramic metal halide and LED.

Wall Washing and Grazing

Wall washing (see **Figure 3-8**) is used to articulate texture and also enables room surfaces to be used to increase light levels and perception of brightness. It entails evenly lighting a wall from top to bottom in a smooth graded wash, calling for a uniform distribution of intensity across its surface. Shadows on the surface are eliminated, hiding imperfections and flattening its visual appearance. Virtually any light source can be used, although linear sources can be very effective for large surfaces. Typical wall wash luminaires include track lighting and ceiling-mounted luminaires placed at a constant distance from the wall and each other. The luminaires must be placed at a sufficient distance from the wall to produce the washing effect—in some applications, 2.5 to 3 ft. might be suitable. Luminaire quality and placement are critical

Figure 3-7. Downlights. Madison Area Technical College installed 400 EN-ERGY STAR-qualified LED luminaires across their campus, including the cafeteria shown here. Photo courtesy of Cree, Inc.

to producing uniform lighting.

Because of the Inverse Square Law, wall washing is limited to 8-9 ft. of vertical wall space. Obviously, the bottom zone of the wall near the floor receives the lowest amount of illumination. The floor area is typically obscured by activities and furniture, however, so people are generally focused on the bright upper wall.

Wall grazing (see **Figure 3-8**) is another basic lighting technique that involves placing the luminaires closer to the wall, resulting in a narrower angle, which in turn produces shadows that reveal texture. By moving the luminaire closer to or farther from the wall, the angle of light can be adjusted to make shadowing more or less pronounced and thereby achieve different grazing effects. Linear sources can be used but point sources are generally preferred for strong grazing; LEDs have great potential as an efficient grazing source.

Because grazing reveals texture and is visually arresting, it should be used with surfaces where texture is a sign of beauty, not imperfection. In the case of wall washing, the lighting reveals the spatial form as a luminous backdrop to people and objects. In the case of wall grazing, the wall itself is the focus of attention. As a result, grazing is typically used as accent lighting to beautify strongly textured surfaces such as natural stone and brick as well as artwork such as carvings.

A third technique is also available that combines some of the advantages of wall washing and wall grazing, called a grazing wash (see **Figure 3-9**). This technique involves mounting linear sources in a continuous run around the edge of the ceiling (wall/slot lighting). The lighting equipment can be concealed while the bright wash of light at the edge of the ceiling articulates the ceiling plane, resulting in a visually interesting "floating ceiling" effect. While the angle of light is narrow enough to render interesting architectural details, grazing wash is generally used similarly to wall washing—that is, to reveal the space's dominant boundaries and produce interreflections. It is also particularly useful when wall washing would risk producing reflected glare that would be uncomfortable for people performing tasks near the lighted walls.

Cove Lighting

Cove lighting uses perimeter coves to conceal lighting that projects a pleasing lighting pattern on the ceiling and indirect light distribution, or ambient lighting, into the space. Cove lighting may use linear or

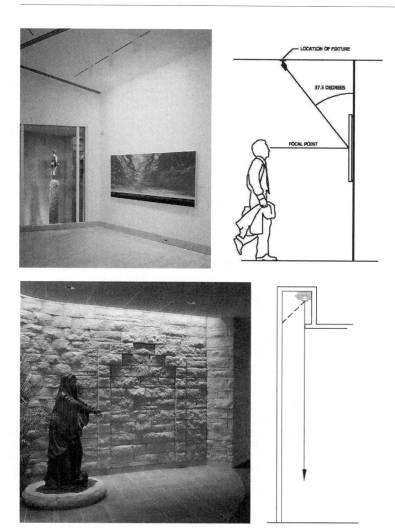

Figure 3-8. Wall washing versus grazing. In the top photo and drawing, we see an example of frontal smooth washing of a wall. The wall is illuminated uniformly from top to bottom in a smooth, graded wash of light, without scalloping, flattening the visual appearance of the wall and hiding blemishes and imperfections. In the bottom photo and drawing, we see an example of wall grazing. The luminaire is mounted closer to the wall, reducing the angle of light and producing shadowing that can reveal the rich texture of materials such as stone and brick. Photography courtesy of Michael Spillers Photography, drawings courtesy of Bruce Yarnell, IALD, MIES, LC of Yarnell Associates, LLC.

Figures 3-9a.

Figure 3-9b. A space illuminated using direct/indirect lighting (a), wall/slot lighting (b) and a combination of the two (c). In the first space, the ceiling and task plane are emphasized. In the second, the walls are emphasized while providing a visually interesting "floating ceiling" effect. By combining the two approaches, the complete spatial envelope is dramatically revealed. Images courtesy of Litecontrol.

Figure 3-9c.

Figure 3-10. Cove lighting. Photo courtesy of Cooper Lighting.

Figures 3-11. In this uplighting application, cove lighting is turned upside down for a fresh approach. Images courtesy of Yarnell Associates.

point sources: While linear sources are economical, make sure that the luminaires are placed properly to create a seamless pattern by avoiding socket shadows. Shadows can be avoided if the top of the lamp is level with the top of the cove.

Uplighting

Uplighting places light on objects or surfaces above a luminaire that aims light in an upward direction (see **Figure 3-11**). Its effect is either very desirable or undesirable because it is unusual; effects range from the intimate to eerie. Uplighting with a simple candle can produce intimacy in a restaurant. A lot of landscape lighting includes uplighting to accentuate bushes and trees. This directional technique is also used to highlight architectural surfaces and details indoors and out, and ceilings and walls in indoor spaces.

Silhouetting

While rarely employed in commercial buildings, light can be used to silhouette objects to create a striking display of a sculpture, logo or other object or architectural feature. Light striking the front of the object is softened or eliminated while light strikes the back of the object, rendering it in silhouette. The backlight could be intense or diffused depending on the clarity needed for the silhouette.

Sparkle and Glitter

Sparkle and glitter, or tiny points of high brightness, can create visual interest and contribute to an atmosphere of elegance. As with other lighting techniques, this technique must make sense with the environment—sparkle and glitter would look great in a high-end restaurant (sparkle on silverware in a restaurant, a chandelier, a fiber-optic or LED "starry sky"), but would be confusing and distracting in a fast-food restaurant.

Architectural Integration

Architectural lighting may involve using lighting to accentuate architecture and space features rather than directly participate as architecture or a space feature. In other cases, lighting can make a direct statement either as a focal point, supporting architectural element or simply providing visual cues about the space, its owner and its occupants (see **Figure 3-12** and **Figure 3-13**). The level of architectural integration that is required—whether the lighting equipment must blend or entertain— will determine what luminaires are appropriate for the project.

When luminaires must "disappear," consider lighting designed for mounting and operation in architectural features such as coves, ceiling and wall slots, soffits and valances (see **Figure 3-14**). If using recessed downlights, consider specifying luminaires with small apertures. If luminaires are exposed but should be visually subordinate to the design, blend them into the background using equipment with a very neutral appearance.

The size, finish and mounting height of the luminaires must be coordinated for the design to visually make sense. The luminaires and their placement should respect the scale of the architecture. Similarly, the layout of luminaires must also make sense. Regular grids of luminaires connote a businesslike atmosphere, while random patterns of luminaires, as long as they are not visually confusing, can connote

Figure 3-12. Luminaires that stand out can be used to grab attention and contribute directly to the space appearance. Photo courtesy of Énergie.

Figure 3-13. The appearance of luminaires can contribute to the architecture and aesthetics of a space, conveying information about the space, its owner and its occupants. Photo courtesy of Peerless Lighting.

Figure 3-14. Architectural features such as coves, ceiling and wall slots, soffits and valances provide opportunities to conceal lighting. Photo courtesy of Litecontrol.

a more casual atmosphere. Rows of linear luminaires can appear to visually lengthen corridors, while illuminating a corridor's walls using recessed wall accent lighting can visually shorten it. Regardless of approach, it is vital to avoid visual clutter and confusing cues. Of course, the lighting layout must be coordinated with the architecture to ensure it can be installed as it appears on the reflected ceiling plan.

Overall, remember that light is the medium through which a space's users interact with the architecture: Beautify spaces with light. What do you wish people to see?

Light and Shadow

The interaction of light and shadow can have functional and aesthetic impacts on faces, objects, surfaces and spaces. Shadows can impact task visibility and influence perception of spaces by what light reveals and shadows conceal. It can change the visual appearance of objects and surfaces by accentuating or washing out details. It can make faces appear old or young, attractive or ugly, friendly or sinister (see **Figure 3-15**).

Figures 3-15. Lighting can affect how faces and objects are perceived; in the case of facial modeling, lighting can play a significant role in the success or failure of social interactions. Here we see a statue of a woman photographed under four different lighting conditions: diffuse lighting (top left, diffuse lighting with sidelight (top right), downlighting only (bottom left), and strong uplighting (bottom right). With diffuse lighting only, her face is bland, lacking definition. With downlighting only, her face is rendered harshly—every line and wrinkle is accentuated by shadows. With uplighting, she looks sinister. With diffuse lighting combined with sidelight, a good balance is achieved: There is good facial definition without undesirable shadowing. Photos courtesy of Naomi J. Miller, FIES, FIALD, LC.

In an article for Architectural Lighting Magazine, lighting designer and author Gary Gordon, IES, IALD points out that light and dark need not be antagonists. Instead, he writes, "They are counterparts, like the *yin* and *yang* of Chinese cosmology that combine to produce all that comes to be. Without shade or darkness, light loses much of its meaning; patterns of light and shade render the prominences of surfaces and objects in the visual field."

Typically, large and pronounced shadows are undesirable in the built environment and should be avoided in spaces where critical tasks are performed and face to face communication is important. But if shadowing were to be completely eliminated, texture would be washed out and people and objects would appear visually flat and less interesting. Some shadowing is typically desirable for depth and interest. Shadows can reveal details in a complex visual task, enhance the aesthetic of textured surfaces such as brick walls, aid in facial recognition and social communication, enhance the presentation of sculpture, and even enable interesting patterns to be projected onto walls and ceilings by shining a light behind an object such as a plant. The lighting designer should consider the directionality of light sources in the space, and associated intensities, to strike the right balance between light and shadow to render faces and objects in their best light.

The Role of Surfaces in Lighting Efficiency

Surfaces and objects in a space may be considered an extension of the lighting system. If light is not absorbed, it is reflected or transmitted with reduced intensity to other surfaces and objects, where it is absorbed, reflected and/or transmitted. All surfaces and objects in a space that reflect or transmit light can be considered an extension of the luminaires or, rather, a secondary light source with its own particular and notable characteristics.

Surfaces with lighter finishes have higher reflectances and therefore promote interreflections of light in the space (see **Table 3-2**) that soften shadows, reduce contrast, increase uniformity and raise light levels. Using indirect lighting, a wall or ceiling can be the major source of illumination in a space. As a result, materials, finishes, furniture, etc. are important considerations in lighting design, as lighter finishes can reinforce luminaire output through reflection and darker finishes can negate it through absorption. Room surface finishes, therefore, can play a significant role in maximizing the efficiency of a lighting system because lighter finishes can produce relatively higher light levels and perceptions of brightness. In fact, in an existing space with a large area of dark surfaces, light levels and visual comfort can be improved almost immediately by simply repainting those surfaces a lighter color to increase reflectance—and thereby improve *coefficient of utilization*, an expression of efficiency of a luminaire in a given application and an important factor in one method of design light level calculations (see **Figure 3-16**).

Table 3-2. ASHRAE-recommended surface reflectances. Source: ASHRAE Advanced Energy Design Guides.

Office	K-12 School	Small Retail	Small Hospital and Healthcare	Warehouse and Self-Storage
Ceiling: ≥80% (90% if indirect lighting) Walls: ≥70% (same for >2.5-ft. vertical partitions)	Ceiling: 70% (preferred 80-90%) Walls: 50% Floor: 20%	Ceiling: 80% (80+% if daylight zone) Wall: 50% (70+% if daylight zone) Floor: 20%	Ceiling: 85% (direct lighting) and at least 90% (indirect and/or daylighting) Walls: 50% (70% for walls adjacent to daylight apertures) Floor: 20%	Ceiling: 80% Walls: 30% Product/Floor: 20%

| RCC | 80 | | | | 70 | | | | 50 | | | 30 | | | 10 | | | 0 |
RW	70	50	30	10	70	50	30	10	50	30	10	50	30	10	50	30	10	0
RCR																		
0	.86	.86	.86	.86	.73	.73	.73	.73	.50	.50	.50	.29	.29	.29	.09	.09	.09	.00
1	.78	.74	.71	.68	.66	.63	.61	.58	.43	.42	.40	.25	.24	.24	.08	.08	.08	.00
2	.71	.64	.59	.55	.60	.55	.51	.48	.38	.35	.33	.22	.21	.19	.07	.07	.06	.00
3	.64	.57	.51	.46	.55	.49	.44	.40	.33	.30	.28	.19	.18	.17	.06	.06	.05	.00
4	.59	.50	.43	.39	.50	.43	.38	.33	.29	.26	.24	.17	.15	.14	.05	.05	.05	.00
5	.54	.43	.38	.33	.46	.38	.33	.28	.26	.23	.20	.15	.13	.12	.05	.04	.04	.00
6	.49	.39	.33	.28	.42	.34	.28	.24	.23	.20	.17	.14	.12	.10	.04	.04	.03	.00
7	.45	.35	.29	.24	.38	.30	.25	.21	.21	.17	.15	.12	.10	.09	.04	.03	.03	.00
8	.42	.32	.25	.21	.35	.27	.22	.18	.19	.15	.13	.11	.09	.08	.04	.03	.03	.00
9	.39	.28	.22	.18	.33	.25	.19	.16	.17	.14	.11	.10	.08	.07	.03	.03	.02	.00
10	.36	.26	.20	.16	.31	.22	.17	.14	.16	.12	.10	.09	.07	.06	.03	.02	.02	.00

Floor Cavity Reflectance .20

Figure 3-16. This coefficient of utilization table is published for a given indirect general lighting product considered for installation in a room with a 20 percent floor reflectance, 30 percent ceiling reflectance and 50 percent wall reflectance. The room cavity, defined as the volume of space between the luminaires and the workplane, is 5 ft. high x 15 ft. long by 15 ft. wide, so the room cavity ratio, using the formula RCR = [5 x H x (L + W)] ÷ (L x W), would be 3. The dimensions are fixed and cannot be changed, so looking up an RCR of 3 in the below table, we get a CU of 0.19. If Required Lumens = [Maintained Illumination (fc) x Area (sq.ft.)] ÷ (Ballast Factor x CU x Light Loss Factors), and our maintained light level is 35 footcandles, ballast factor is 0.88 and the overall light loss factor is 0.75, then 63,000 lumens are required for this 225-sq.ft. space.

Similarly, consider the height of room partitions in open office plans and similar spaces. Taller partitions provide privacy for occupants. Lower partition heights, however, enhance daylight penetration, uniformity and access to a view, while minimizing shadows.

Now suppose lighter finishes are used that raise ceiling reflectance to 80 percent—this is, after all, a 100 percent uplight luminaire, so 80-90 percent is recommended—with wall reflectance increased to 70 percent for non-window walls and 80 percent for window walls (to reduce contrast between bright windows and their surround). Looking at the CU table for the luminaire, CU is now 0.64. And our lumen requirement drops to about 19,000 lumens—an instant potential 70 percent savings on capital and operating costs. As shown, controlling surface reflectances can impact overall application efficiency, particularly when indirect luminaires are being considered.

In the Spotlight: Spacing Criteria for Direct Luminaires

Luminaire spacing is a critical design decision for direct luminaires such as troffers and downlights. If luminaires are spaced too close together, unnecessary capital may be invested in the system. If spaced too far apart, the space may appear unevenly illuminated, with dark spots. Direct luminaires should be placed within the manufacturer's recommending spacing criterion (SC) or spacing-to-mounting height ratios, typically published in product photometric reports or catalog sheets.

Recommended maximum spacing from the center of each luminaire to the next is typically calculated as SC x Mounting Height. So if SC = 1.2 for downlights mounted on an 8.5-ft. ceiling, the recommended maximum spacing between the luminaires would be 10.2 ft. This should provide a uniform pattern of light with a little overlap of each luminaire's pattern. If the luminaire is a rectangular-shaped troffer, two spacing criteria may be published, one for the maximum recommended spacing between luminaires 90°, or across the length of the luminaires, and one for 0°, or along the length of the luminaires. As one might expect, the spacing criteria at 90° is often greater than the spacing at 0°.

Luminaires should not be placed too far from walls; it is sometimes recommended that the distance between walls and adjacent luminaires should not be larger than one-half of the spacing based on the manufacturers recommended SC. This places some illumination on upper walls, making the space appear brighter and more spacious. Alternately, dedicated wall lighting can be specified, such as wall washers or wall/ slot systems, or indirect general lighting.

Note that most decorative luminaires and accent lighting are not subject to spacing criteria. Decorative luminaires should be mounted where appropriate. For accent lighting, the manufacturer may offer guidelines regarding the distance between the luminaire and the object based on the ceiling height, enabling the designer to ensure the object is uniformly lighted. Similarly, pendant indirect luminaires also have no spacing criteria, although manufacturers provide recommendations on spacing that can be used to achieve good uniformity.

IMPACT OF DAYLIGHT

While not an electric light source, daylight is an important layer to consider in a lighting design in any space that has daylight apertures such as windows or skylights. In its simplest definition, daylighting is the use of daylight as the primary source of illumination in a space. Daylighting is also related to providing access to a view, optimizing quality of the visual environment, and reducing whole building energy consumption. When designing a lighting system, one must consider where daylight is allowed to enter the space and in what intensity, how it is controlled, and how the electric lighting and daylighting will be integrated.

Access to a View

While view windows can be potential sources of glare (that can be addressed with good design and technologies such as micro-blinds), a majority of people want them.

In windowless offices, people tend to experience stress and feelings of being cooped up, feelings that can be alleviated by providing access to a view. Jay Appleton, in *The Experience of Landscape* (1975), theorized that humans evolved a preference for environments that are safe locations from which they can view and survey the surrounding environment (the prospect-refuge theory). Occasionally viewing distant outside objects relaxes the eye muscles and eases discomfort. People also tend to prefer a connection with nature and a sense of time. In short, they want a view.

In a 2003 Heschong Mahone Group study of 100 workers in a call center, workers with the best possible view (and all other things being equal) processed calls 6-12 percent faster. In a related study of 200 office workers, those with ample view performed 10-25 percent better on a variety of cognitive tests versus workers with no view. Those workers with no view self-reported greater fatigue throughout the workweek. This research suggests that a view can be as valuable as having daylight in a space.

Benefits of Daylight

Daylighting can impact people and spaces by providing sensory availability, connection to nature, time/weather information, full-spectrum light, modeling and an indirect component of light producing wall- and ceiling-washing effects, which can provide a more pleasant and comfortable visual environment. Many of these benefits boil down to simple mental stimulation due to moderate changes in the environment, so long as these changes are meaningful and patterned, which research indicates is beneficial to workers in monotonous, uniform office environments.

Figure 3-17. Automatic daylight control devices, such as the shading system shown here, enable view and daylight to be maximized while minimizing glare to the interior. Photo courtesy of MechoShade Systems, Inc.

The impact can be dramatic, as indicated by numerous studies over the past 50 years. Various Heschong Mahone studies, for example, discovered an increase in sales as high as 40 percent in retail stores with skylights versus those without any daylighting, and a 21 percent improvement in learning rates (one study) and 7-18 percent higher test scores (another study) in school classrooms with daylighting.

In retrofit applications, adding daylighting is constrained by the existing building design but there may be opportunities to incorporate solutions such as tubular daylighting devices that "pipe" daylight into interior spaces.

Integration with Electric Lighting

Daylighting should be controlled to avoid unwanted glare and heat gain, while the lighting system should be designed to properly integrate daylight.

To integrate daylight with the electric lighting, the luminaires should emit light on the same surfaces and in the same direction as the daylight apertures. If daylight is placed on high walls and ceilings, then the electric lighting system should place light on these surfaces as well. In deeper spaces where daylight does not penetrate to the rear areas, consider wall washing on the rear wall to prevent excessive contrasts. To prevent excessive contrasts between daylight apertures and surrounding wall or ceiling, consider lighter finishes and placing light on those surfaces.

Besides distribution, the light source itself should be specified with daylight in mind: Diffuse sources such as fluorescent complement daylight's diffuse characteristics. Consider cooler (4000K to 6000K) light sources in spaces occupied mainly during the day to match the color temperature of daylight; if cooler light sources are unwelcome, consider a neutral (3500K to 4100K) source.

Additionally, if daylight is a primary source of illumination, consider adding lighting controls that automatically dim or turn OFF the lights in response to daylight contribution to task light levels. (Alternately or in addition to this, consider manual controls enabling users to reduce or turn OFF the lights if ample daylight is present.) Switching is generally considered more suitable for spaces where users do not perform critical tasks, such as lobbies and atria, while dimming is generally considered more suitable for spaces where they do, such as offices and classrooms. When daylight harvesting control (controls that automatically reduce lighting in response to available daylight, saving energy) will be employed, the lighting system ideally will be designed so that circuiting aligns with daylight patterns, and enable occupants to adjust or override the automatic control.

In the Spotlight: Daylight Harvesting at NRG Systems

In 2005, Vermont-based manufacturer NRG Systems built a new headquarters carefully crafted to reflect the company's commitment to the environment, the community and its employees. The 46,550-sq.-ft. facility, which includes office, manufacturing and warehouse space for company-produced wind monitoring equipment, was designed to minimize environmental impacts and maximize energy conservation. One of only a handful of manufacturing facilities to earn a LEED Gold certification, it serves as an example of daylight harvesting done right.

In this office space, daylight enters at a high point via a louvered daylight aperture, the indirect luminaires are circuited parallel to the daylight source, and the rows of luminaires are dimmed according to daylight contribution to the space. Continual energy monitoring has proven that the combination of daylight harvesting and occupancy sensors used in the building presents actual daytime lighting energy use that is more than 40 percent less than the connected load, even in winter months.

Figure 3-18. Daylighting at NRG Systems. Photo courtesy of WattStopper.

ENERGY-EFFICIENT LIGHTING DESIGN IDEAS

Lighting and energy researchers are continuously exploring new methods to save energy while supporting lighting quality. This section identifies several notable efforts.

Education: Integrated Classroom Lighting System

In many schools, lighting eats up 30-40 percent of utility costs. As energy codes become more restrictive, can lighting satisfy the demands of the modern classroom, with horizontal and vertical task planes, computers and A/V equipment?

To test one approach, the California Energy Commission (CEC) and the New York State Energy Research and Development Authority (NYSERDA) engaged studies of a new Integrated Classroom Lighting System (ICLS) developed by manufacturer Finelite.

ICLS includes two rows of direct/indirect linear fluorescent pendants, mounted parallel to the windows and spaced about 15 ft. apart, with a wallwasher illuminating the main teaching board (see **Figure 3-19**). Each luminaire includes three high-performance (3100-lumen) T8 lamps: two outboard lamps producing uplight and downlight, and a separately ballasted inboard lamp producing downlight. Both the inboard lamp and outboard lamps cannot be on at the same time, resulting in immediate energy savings.

The luminaires are integrated into a control system featuring a ceiling-mounted dual-technology occupancy sensor placed between the rows of pendants, a master switch at the door and a "teacher control center" located near the main teaching board, which features:

- a "Whiteboard" switch that turns the wallwashing luminaire mounted on the main teaching board ON and OFF;

- a "General/AV Mode" enabling the teacher to switch between General mode (downlight OFF, uplight/downlight ON) and A/V (and reading) Mode (downlight ON, uplight/downlight OFF; and

- a "Quiet Time" switch that overrides the occupancy sensor for one hour, keeping the light on during long periods of occupied non-movement such as standardized testing.

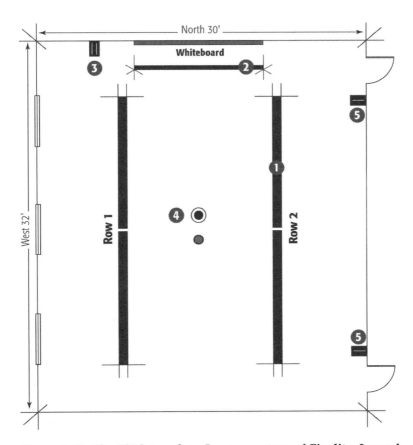

Figure 3-19. The ICLS template. Image courtesy of Finelite. Legend:

1) Two rows of two-scene direct/indirect luminaires mounted perpendicular to the main teaching wall (parallel to window wall) and spaced 14-15 ft. apart.
2) A dedicated luminaire is used to illuminate the whiteboard on the main teaching wall.
3) Teacher control is placed at the front of the classroom. For easy teacher access place controls within 6 in. of the whiteboard.
4) Sensors are placed in the center of the classroom. Sensors always include occupancy and daylight harvesting is added where appropriate.
5) A master ON/OFF switch is by every door to the classroom.

Optionally, teachers are also able to access A/V Dimming Mode, which allows them to turn ON and then dim the inboard lamp providing the downlight component, and which of course requires a dimmable ballast. As another option is that a photosensor can be added, adjusting light output based on daylight availability.

Results of the NYSERDA study include:

• Because all three lamps cannot be ON at the same time, the maximum lighting power density is capped at about 0.8W/sq.ft.

Figure 3-20. Hunter High classroom with the lights ON in General Mode. Photo courtesy of Finelite.

• The switching controls reduced average lighting power density to
  0.73W/sq.ft., about half of ASHRAE 90.1 2004/2007 and about 40
  percent less than Title 24-2005.

• The system was installed for $1.83-$2.29/sq.ft.; options such as
  daylight switching and a third luminaire row to increase uniformity add to the cost).

Many luminaire manufacturers now offer solutions based on this
template.

**Figure 3-21. Hunter High classroom with the lights ON in A/V Mode. Photo
courtesy of Finelite.**

Office: ASHRAE Advanced Energy Design Guides

ASHRAE, in collaboration with government and other industry organizations, has published a series of free Advanced Energy Design Guides (AEDG) at www.ashrae.org that provide a prescriptive path for beating ASHRAE 90.1-1999, the national energy standard until the end of 2010, by 30 percent in small office, small retail, K-12 school, warehouse and small hospital and healthcare buildings.

In this section, we will focus on AEDG recommendations for small office buildings (<20,000 sq.ft). Regarding lighting, this means reducing lighting power density (LPD) from 1.3W/sq.ft. to 0.9W/sq.ft. This is just 0.1W/sq.ft. less than 90.1-2004's 1W/sq.ft. limit. It is achievable using today's technology and design practices.

In an office, it is assumed that the target ambient light level is 30 footcandles and the task light level on the desktop is 50 fc. The Guide recognizes distributing light on walls and ceiling as good design practice, and therefore recommends one- or two-lamped direct/indirect luminaires mounted in continuous rows parallel to the windows, with some luminaires close to walls or installed with supplemental wall-wash luminaires.

To increase light levels with indirect lighting, the ceiling should have a 90 percent reflectance, if possible, or at least 80 percent; most ceiling tiles have a 70-80 percent ceiling reflectance, while some manufacturers offer high-reflectance ceiling tiles. Walls and >2.5-ft. vertical partitions should have a 70+ reflectance finish value.

The Guide recommends high-performance T8 lamps, although T5 and T5HO lamps may be used as long as the luminaire does not have an open bottom. High-performance T8 lamps are lamps that produce more light (3100+ lumens) for the same wattage, paired with low-output (0.77 ballast factor) ballasts to save energy, or lamps that produce similar light output (2750-2850 lumens) for less wattage (28W or 30W). T8 lamps should be operated by instant-start electronic ballasts, which provide higher energy savings, lower cost and parallel operation (if one lamp burns out, its companion will continue operating normally), although programmed-start ballasts are recommended to extend lamp life if the lamps are frequently switched by occupancy sensors.

Which, by the way, the Guide recommends: Occupancy sensors, set to medium to high sensitivity and a 15-minute time delay, are recommended in open and private offices, with manual-ON, automatic-OFF operation recommended for private offices. If occupancy sensors do not make sense for an open-plan office space, consider time-scheduling instead. Multi-level switching will generate energy savings while private offices are occupied. Similarly, undercabinet luminaires should include T8 lamps and a two-step ballast with the lower output level hardwired to allow bi-level switching control—or dispense with undercabinet luminaires altogether and use articulated compact fluorescent task lights plugged into strips controlled by local occupancy sensors. Finally, utilize daylight dimming controls for direct/indirect luminaires placed within 12 ft. of windows on the north and south perimeter zones, but note that 28W and 30W T8 lamps recommended by the Guide currently offer limited compatibility with dimming ballasts; additionally, the photosensor should include a minimum five-minute time delay to avoid cycling.

Overall, the Guide provides a prescriptive lighting path to exceed 90.1-1999/2001 by 30 percent and 90.1-2004 by about 10 percent.

Lighting recommendations for the other building types covered by the Advanced Energy Design Guides are shown in Table 3-3. See the individual Design Guide for specific guidance, downloadable free at www.ashrae.org.

Office: Task/Ambient lighting design with LED task lighting

Task/ambient illumination is a lighting design approach that combines a low-level direct/indirect general lighting layer providing ambient illumination and a more intensive task lighting layer providing sufficient light levels to complete critical visual tasks. The use of indirect lighting makes the space appear visually brighter and more spacious, while the focus of lumens and watts at the supplementary task layer improves efficiency.

The reasoning is sound: Any time a light source is located closer to the task, a gain in efficiency will result (per the Inverse Square Law) as long as the result does not overlight the space or use relatively inefficient luminaires. Estimates of potential savings derived via installation of single-lamped indirect pendants and task lighting in typical office workspaces range from 20 percent, according to the Midwest Energy Efficiency Alliance, to 33 percent, according to American Council for an Energy-Efficient Economy (ACEEE).

According to ACEEE, the main application for this approach is private offices and some meeting spaces, although some application could reasonably be found in open office plans and other applications. In some applications, a task/ambient approach can be particularly

Table 3-3. Summary of key lighting recommendations for major building types in ASHRAE Advanced Energy Design Guides (other than small office buildings).

	K-12 School	Small Retail	Small Hospital and Healthcare	Warehouse and Self-Storage
Interior Finishes	Ceiling: 70% (preferred 80-90%) Walls: 50% Floor: 20%	Ceiling: 80% (80+% if daylight zone) Wall: 50% (70+% if daylight zone) Floor: 20%	Ceiling: 85% (direct lighting) and at least 90% (indirect and/or daylighting) Walls: 50% (70% for walls adjacent to daylight apertures) Floor: 20%	Ceiling: 80% Walls: 30% Product/Floor: 20%
Special notes on luminaire distribution		Extensive use of totally indirect or recessed direct/indirect luminaires may not achieve desired light levels for AEDG-specified power density and may create brightness/contrast problems		Use luminaires with aisle distribution to drive light to lower shelves in high-ceiling areas; use tandem 8-ft. luminaires instead of 4-ft. luminaires to spread light across aisle length
CRI	≥80 for ambient lighting	≥80 CRI for ambient/accent lighting	≥80 for ambient lighting	Not considered important
Color Temperature				4100K or 5000K
Glare	If using T5 lighting for low-ceiling applications, use special shielding to avoid glare; properly shield high-bay luminaires and skylights in gyms to avoid direct glare during sports where athletes must track a moving ball	Luminaires should be properly shielded, avoiding highly specular louvers, cones or reflectors visible to occupants; avoid direct lighting of specular surfaces	Patients lying in bed should not be exposed to direct glare from ceiling luminaires or windows, avoid reflected glare on monitor screens, ensure operating rooms are glare-free	Luminaires should be properly shielded to reduce direct view of lamps, avoid clear lens and un-louvered luminaires; use more or longer luminaires with fewer lamps to minimize bright points on ceiling
Technologies Favored	High-performance T8, T5, T5HO linear fluorescent, CFL, metal halide, LED exit signs	High-performance T8, T5, T5HO linear fluorescent, CFL, ceramic metal halide, halogen IR, LED, LED exit signs	High-performance T8, T5, T5HO linear fluorescent, CFL, metal halide, LED exit signs	High-performance T8, T5HO, CFL, metal halide, LED exit signs
Lighting Controls	General and A/V mode, dimming optional, occupancy sensor, daylight harvesting for classrooms	Occupancy sensors (non-sales areas), daylight harvesting, time separate circuiting of general and accent and display case lighting with time sweep control	Occupancy sensors, bilevel switching, wall switches and dimmers (patient care spaces), patient controls, time sweep, daylight harvesting	Occupancy sensors, daylight harvesting

advantageous, such as spaces where high, dark or articulated ceilings or required use of inefficient general lighting (such as for aesthetic reasons) can reduce efficiency in delivery of light to the task. It is also particularly advantageous when primary tasks in the space require very high light levels and when high- and low-level tasks share the same space, particularly where the ratio of circulation space is significantly greater than high-level task space.

The California Lighting Technology Center (CLTC), Public Interest Energy Research (PIER) program and manufacturer Finelite teamed up to produce an integrated office lighting system combining a task/ambient design with flexible LED task lights.

First, the researchers studied offices to find the "typical open office": rows of cubicles often including undercabinet linear fluorescent task light, with general lighting usually provided by a regular array of lensed or parabolic troffers.

The researchers conceived a lighting system that would utilize direct/indirect illumination to improve visual comfort, general lighting

Figure 3-22. The Personal Lighting System improves the quality of task lighting while contributing to dramatic reductions in energy consumption in open office workstations. Photo courtesy of Finelite.

providing a low ambient light level of 25-30 footcandles, and supplementary task lighting providing 40-50 footcandles of no-glare illumination on the task plane with enough light on partition walls to eliminate shadows.

The specially designed task lighting system consists of LED undercabinet and freestanding task lights in 3W, 6W and 9W units; any combination of units can be used as long as the total load is no larger than 21W. Plug-and-play connections enable the task lights to be easily installed and relocated. The units produce suitable desktop and vertical light levels without creating offensive glare. And an occupancy sensor connected to the power supply provides automatic shutoff of the task lighting in the cubicle.

Michael Siminovitch, director of the CLTC, says this approach has demonstrated suitable light levels and lighting quality at a power density of 0.5-0.7W/sq.ft., which is 36-55 percent lower than the ASHRAE 90.1-2004 energy standard. This translates to 40-50 percent energy savings compared to the average California office. Meanwhile, the occupancy sensor generated an additional 20-30 percent energy savings.

Office and Retail: DOE Commercial Lighting Solutions Web tool

As energy codes exhaust the potential efficiencies of lamps and ballasts, policy makers are beginning to look to lighting controls, daylighting and innovative design for deep energy savings. As a result, lighting is getting more complex. And as it gets more complex, the risk of producing inadequate lighting increases. While many designers are highly proficient with lighting, not many people know how to do good lighting on a very low energy budget outside the industry's top designers.

To solve this problem, DOE studied the question of how to encourage adoption of advanced lighting technologies and design practices by making them available to everybody in the lighting specification community, not just the leaders in the field. DOE understood that people need not only education, but guidance—guidance that is accessible, easily understandable and, most important, actionable.

The result is the Commercial Lighting Solutions program, a web tool providing customizable lighting templates designed to generate up to 30+ percent lighting energy savings compared to ASHRAE 90.1-2004/2007, without sacrificing lighting quality. It is specifically designed for people who make lighting decisions but are not necessarily lighting experts. Contractors, designers, distributors and owners can use these templates to achieve the latest, energy-efficient solutions—while providing good lighting.

CLS launched at LIGHTFAIR 2009 with a retail version. An expansion for office lighting was subsequently fast-tracked to support large public spending projects, and was launched at LIGHTFAIR 2010. The stakes are potentially high, as $5 billion has been earmarked for Federal building upgrades alone, and an estimated $1 billion is being spent on lighting. While the CLS for Office tool will not be required, the General Services Administration (GSA) may encourage its use by Federal facility managers responsible for upgrading their buildings.

After registering, the user provides some basic information about the building, such as its operating hours, location and applicable energy code. For office buildings, a drawing then appears showing six types of office spaces: private offices, open offices, open office perimeter, corridors, conference rooms and reception. The user selects a space and enters some information about it, such as the ceiling height, total area and whether windows or skylights provide daylight. If daylight is present, the user then enters additional information, such as amount of glazing, visible transmittance and whether there are light shelves installed.

Lighting choices are shown to the user that match the space characteristics. Overall, some 40 major lighting and control combinations are available, with the lighting templates designed by lighting design firm Horton Lees Brogden. In open office spaces, for example, the user can choose lighting layouts including 8-ft. multilamp direct/indirect luminaires, 12-ft. single-lamp direct/indirect luminaires, or recessed non-planar lensed luminaires, with lighting power densities up to 40 percent lower than ASHRAE 90.1-2004/2007 using the Space by Space Method. Suppose we choose 12-ft. single-lamp direct/indirect pendants. The luminaires place light on the walls and ceilings, making the space appear larger and brighter with improved uniformity, while adjustable desktop task lights provide supplementary lighting and give users personal control and increased satisfaction. Average maintained horizontal work surface light levels range from 30-35 footcandles ambient and 45-75 footcandles on paper tasks; ballasts can be specified with lower or higher ballast factors to tune light levels to the project. The lamps have a neutral color appearance and good color rendering, with a color rendering index rating of 85 or higher.

Once the user selects the lighting solution, control options are presented (this book's author worked with the industry to develop the control templates). For open offices, the control options range in capabilities from networked vacancy sensors to adding daylight harvesting to performing multiple control strategies via a highly integrated, individually addressable, centrally controllable, digital lighting control system. Depending on the option selected, energy savings can increase significantly. Each control option includes a zonal map revealing the control strategy, generic performance specifications (based in part on energy code requirements and GSA specifications) and special notes on installation and commissioning, making it easy to implement.

The CLS web tool is available free to the public and can be accessed at www.lightingsolutions.energy.gov.

Demand-responsive Lighting

Satisfying peak demand can be very expensive for utilities, which pass this cost on to their customers in the form of time-variable pricing or demand charges.

Figure 3-23. CLS provides design templates enabling lighting specifiers to achieve 30 percent more energy savings than the ASHRAE 90.1-2004 energy standard without sacrificing lighting quality. Image courtesy of DOE.

Utilities, therefore, share a common interest with their customers to reduce peak demand. This is because shaving the peak enables them to satisfy customer demand while avoiding the high cost of building new capacity or having to buy very expensive power from other markets during an emergency or demand spike. Besides charging more for power used during peak demand periods, a number of utilities offer financial incentives to building owners to curtail load on request, usually during an emergency grid event.

To reduce peak demand, we can turn equipment OFF, turn it down or use it more efficiently. Lighting, while being a major energy consumer in a typical commercial building, at first glance has a small role to play: While we can use it more efficiently, it is difficult to turn OFF in routinely occupied spaces for obvious reasons, and cannot be turned down in many spaces without installing dimmable ballasts.

But suppose we do just that: replace every fluorescent ballast in a commercial building with a dimmable ballast.

A number of solutions are available from manufacturers. They include ballasts that communicate via low-voltage wiring and therefore can be integrated into a lighting control system that leverages dimming for other energy-saving purposes. They also include line-voltage ballasts, ideal for retrofit, designed to provide continuous or stepped dimming.

How low could we go with light levels during peak demand events?

Guy Newsham, PhD, senior research officer at the National Research Council Canada—Institute for Research in Construction (NRC-IRC), led a group of researchers to study this question that is at the heart of dimming's potential as a demand-response strategy.

"Our aim was to explore how far and how fast one could dim smoothly from normal levels, and over what period, without incurring undue hardship on occupants," he says.

NRC-IRC conducted two laboratory studies in full-scale office mockups where various dimming scenarios were studied with typical office workers conducting office tasks. The researchers then took the results and used them to design a field study, conducted during the summer, that included an open-plan office with 330 dimmable luminaires and a college campus with 1,850 dimmable luminaires in several buildings. Load shedding was performed during afternoon hours over several days. The rate of dimming spanned one to 30 minutes with dimming reductions up to 40 percent. Occupants were warned that an experiment would be conducted over the summer involving afternoon dimming, but were not told which days.

"In the field study, we attained load sheds of 14 to 23 percent of normal lighting load with no complaints from occupants," says Newsham. "In combination with the lab study data and findings of related work by others, we derived several recommendations. The first stage of demand response should be dimming by amounts that are not even noticed by the large majority of occupants. The second stage of demand response, when more load reduction is required, may involve dimming to light levels that are noticeably lower but are still acceptable to the large majority of occupants."

In Stage 1, dimming can occur rapidly, over as little as 10 seconds, by 20 percent with no daylight, 40 percent with low prevailing daylight, and 60 percent with high prevailing daylight. If dimming occurs slowly, over 30 minutes or more, and with no immediate expectation of diming occurring, levels may drop by 30 percent with no daylight and 60 percent with high prevailing daylight.

In Stage 2, dimming can occur rapidly, over as little as 10 seconds, by 40 percent with no low daylight and 80 percent with high prevailing daylight. If dimming occurs slowly, over 30 minutes or more, and with no immediate expectation of diming occurring, levels may drop by 50 percent with no daylight and 80 percent with high prevailing daylight.

"Nevertheless, it is imperative to recognize that demand-response dimming should only be enacted to alleviate temporary grid stress problems that occur infrequently, and is intended to prevail for a few hours at most, and light levels should be returned to normal levels thereafter. Our studies do not provide support for these lower light levels becoming the 'new normal.'"

The challenge is making the capital investment in dimmable ballasts worthwhile via infrequent reductions in load, particularly if daily load shedding is off the table. Another is liability: Managers may feel at risk if they dim for demand-response purposes and somebody trips and falls and blames low light levels. This may require some leniency in IES recommended lighting practices to accommodate certain temporary deviations.

It is highly likely that dimmable lighting will be a component of demand-responsive commercial buildings in the future, however. According to DOE, about 281 gigawatts of new generating capacity will be needed by 2025 to satisfy growing demand for energy, much of which will be allocated solely to satisfy peak demand.

Chapter 4

Legislation and Codes

Lighting design is beginning to
feel more like regulation engineering than art.
—Kevin Willmorth, lighting consultant

THE LAW ON LIGHTING

In recent years, certain incandescent general-service and reflector lamps, fluorescent general-service lamps and fluorescent magnetic, mercury vapor and probe-start metal halide ballasts have been targeted for elimination by efficiency legislation and regulations. As a result, the retirement of some of lighting's venerable workhorses, already occurring by market preference, is being accelerated in favor of younger, more-efficient, better-performing competitors.

The Energy Policy Act of 2005 expanded ballast regulations put into effect by DOE in 2002. Starting July 2010, with few exceptions, ballast manufacturers are prohibited from producing magnetic ballasts for full-wattage and energy-saving 4- and 8-ft. T12 lamps in new luminaires or even for replacement purposes.

The magnetic ballast will basically be eliminated. And now, with the introduction of new DOE regulations that take effect July 14, 2012, so will a majority of the T12 lamps they operate.

The new DOE rules expand on efficiency rules established by the Energy Policy Act of 1992 by strengthening standards for covered lamp types while also covering 8-ft. T8 lamps, 4-ft. T5 lamps and more wattages of 4-ft. T8 and T12 lamps. The net result is a majority of 4-ft. linear and 2-ft. U-shaped T12 lamps, many 8-ft. T12 and T12HO, and some low-color-rendering 4-ft. T8 lamps will be eliminated. While no longer popular in new construction, an estimated 30 percent of fluorescent 4-ft. lamps sold every year are T12.

In the high-intensity discharge (HID) family, mercury vapor and probe-start metal halide ballasts are being targeted.

Mercury vapor is not specified very often anymore, but still has a large installed base. The Energy Policy Act of 2005 eliminated manufacture and import of these ballasts as of January 1, 2008. The Energy Independence and Security Act of 2007 enacted a few technical corrections allowing use of specialty ballasts. While there may be options to lose the ballast but keep the lamp, it is time for owners of these systems to consider upgrading them to other sources such as metal halide.

Meanwhile, on January 1, 2009, the Energy Independence and Security Act began prohibiting probe-start magnetic ballasts from being sold in new 150-500W metal halide luminaires, with some exceptions, effectively also eliminating most 175-400W probe-start lamps. Alternatives include fluorescent and pulse-start metal halide luminaires, which are more efficient and better performing in almost all respects.

Finally, general-service and reflector incandescent and halogen lamps are covered by the Energy Independence and Security Act and subsequent DOE regulations.

In a nutshell, 40-100W incandescent general-service screw-in lamps will be eliminated starting in 2012-2014 unless they can achieve new efficiency targets. The market is expected to shift to compact fluorescent lamps (CFLs), although there are some halogen screw-in lamps that offer a good alternative, and LED replacement lamps may be introduced that can get a foothold in this enormous market. Additionally, new DOE energy standards for reflector lamps, building on the Energy Policy Act of 1992 and the Energy Independence and Security Act of 2007, will eliminate many 40-205W R, PAR, BR, ER and BPAR lamps (with a diameter >2.5 in.) starting July 14, 2012—again, with some exceptions. The market is expected to shift to halogen infrared (HIR) lamps, low-voltage halogen systems, self-ballasted ceramic metal halide lamps and LED directional lighting.

Of these efficiency regulations, almost all of them target technology that, in some cases, is so obsolete it is surprising the market has not finished them off on its own. For example, probe-start metal halide luminaires were being installed in new buildings that were immediately ripe for retrofit to fluorescent luminaires for up to 50 percent energy savings. For almost all of the target technologies, highly efficient and better-performing substitutes are available.

Fluorescent T12 and incandescent reflector lamps, fluorescent magnetic, mercury vapor and probe-start magnetic metal halide ballasts: They've had a good run, but now it is time to gracefully retire.

Energy Policy Act of 2005 Sets New Ballast Efficiency Standards

With few exceptions, magnetic ballasts designed to operate full-wattage and energy-saving F40T12, F96T12 and F96T12HO lamps are being phased out—even replacement ballasts—starting the clock for every building in the U.S. to switch to T8 systems or T12 electronic ballasts if they have not done so already.

In September 2000, DOE published the Fluorescent Lamp Ballast Energy Conservation Standards (10 CFR, Part 430), which established new minimum ballast efficacy factor (BEF) standards that would go into effect starting in 2005. Ballasts that did not pass the standards would be phased out of production and sale in the United States.

Prior to 2005, many believed that this would spell the end of magnetic ballasts. However, it was determined that while magnetic ballasts that operated F40T12, F96T12 and F96T12HO lamps failed the new BEF standards, ballasts designed to operate energy-saving versions of these lamps were not covered by the regulation.

With support from the National Electrical Manufacturers Association (NEMA), the government revisited ballast efficiency in the Energy Policy of Act of 2005, which was signed into law on August 8, 2005. Provisions in the Act extend the coverage of BEF standards, which resulted in the phased elimination and sale of most magnetic ballasts in new luminaires, including those designed to operate 34W T12 lamps, starting in 2009, and replacement ballasts in 2010.

Table 4-1 displays the ballast efficacy requirements for ballasts operating F40T12, F96T12 and F96T12HO lamps, which went into effect April 1, 2005, and the later requirements for ballasts operating F34T12, F96T12/ES and F96T12HO/ES lamps, which went into effect July 1, 2009.

BEF is calculated: [Ballast Factor * 100) ÷ System Watts]. The higher the number, the more efficient the ballast must be to meet or exceed it. **Table 4-2** displays the two timetables governing the phased withdrawal of ballasts that do not meet the new BEF standards.

Both ballast rules exempt ballasts that:

- are designed for dimming to 50 percent or less of their maximum light output;

- are designed for use with two F96T12HO lamps at ambient temperatures of -20°F and for use in outdoor signs; or

- have a power factor of less than 0.90 and are designed and labeled for use only in residential applications.

A significant element of both the original 2000 ballast rule and the new rule included in the Energy Policy Act of 2005, however, is the gradual elimination of the manufacture and sale of replacement magnetic ballasts for full-wattage and energy-saving F40T12, F96T12 and F96T12HO lamps. Building owners will have to retrofit or relight or gradually replace their magnetic ballasts with higher-efficiency ballasts as they fail, such as T12 electronic ballasts or T8 lamp-ballast systems.

Table 4-1. Ballast efficacy requirements for ballasts operating T12 lamps.

Ballast-Lamp System	Nominal Lamp Wattage (W)	Voltage (V)	Ballast Efficacy Factor (BEF)
(1) F40T12 lamp	40	120/277	2.29
(1) F34T12 lamp	34	120/277	2.61
(2) F40T12 lamps	80	120/277	1.17
(2) F34T12 lamps	68	120/277	1.35
(2) F96T12 lamps	150	120/277	0.63
(2) F96T12/ES lamps	120	120/277	0.77
(2) F96T12HO lamps	220	120/277	0.39
(2) F96T12HO/ES lamps	190	120/277	0.42

Table 4-2. Timetable for phase-out of ballasts that do not meet new efficacy standards.

Action	2005 BEF Standards for Full-Wattage T12 Lamps	2009 BEF Standards for Energy-Saving T12 Lamps
Ballast manufacturers can no longer make ballasts that do not pass the new requirements for use in new luminaires.	April 1, 2005	July 1, 2009
Ballast manufacturers cannot sell ballasts that do not pass the new requirements to U.S. luminaire manufacturers.	July 1, 2005	October 1, 2009
Luminaire manufacturers cannot sell luminaires that include ballasts that do not pass the new requirements.	April 1, 2006	July 1, 2010
Ballast manufacturers cannot manufacture replacement ballasts that do not pass the new requirements.	July 1, 2010	July 1, 2010

Energy Policy Act of 2005 Eliminates Most Mercury Vapor Ballasts

The Energy Policy Act of 2005 prohibited manufacturing and importing mercury vapor ballasts starting January 1, 2008.

The law intended to target standard systems but inadvertently targeted specialty systems as well. As a result, the Energy Independence and Security Act of 2007 updated the Energy Policy Act of 2005 with two technical corrections.

First, mercury vapor lamps are defined as featuring screw-bases, so manufacturers can still produce ballasts to operate uniquely based lamps used in applications such as UV curing and chip manufacturing. Second, if a standard screwbase mercury vapor lamp is to be used in a specialty application, the ballast's label must include the notice, FOR SPECIALTY APPLICATIONS ONLY, NOT FOR GENERAL ILLUMINATION, as well as its intended specialty applications.

Energy Independence and Security Act of 2007 to Phase Out General-Service Incandescents

The Energy Independence and Security Act of 2007 (HR6) includes few provisions directly related to lighting, but one of them will have a major impact on lighting—the phased elimination of many common types of incandescent lamps starting in 2012.

According to the California Energy Commission, about 2 billion incandescent lamps are sold each year (2005). The Act will turn this market on its head, putting more than 4 billion general-service incandescent light sockets up for grabs.

The Act defines a general service incandescent as an incandescent or halogen lamp, standard or "modified spectrum" and intended for general service applications, featuring a medium-screw base, producing 310-2600 lumens of light output and capable of operating within 110-130V.

In a nutshell, the Act requires all general-service lamps to have a minimum CRI and general service incandescent lamps to operate at a minimum efficiency.

What are the rules?

With phased implementation beginning January 1, 2012, general-service incandescent lamps must become about 30 percent more efficient, or face elimination, as shown in **Tables 4-3** and **4-4**.

Starting January 1, 2012, all general-service lamps, including CFLs, LED, incandescent and halogen, must have a minimum color rendering index (CRI) rating of 80 if a standard general-service lamp and 75 if a

Table 4-3. Energy standards for general-service incandescent lamps.

Rated Lumen Ranges	Maximum Rate Wattage	Minimum Rate Lifetime	Effective Date
1490-2600	72	1,000 hours	1/1/2012
1050-1489	53	1,000 hours	1/1/2013
750-1049	43	1,000 hours	1/1/2014
310-749	29	1,000 hours	1/1/2014

Table 4-4. Energy standards for modified-spectrum general-service incandescent lamps.

Rated Lumen Ranges	Maximum Rate Wattage	Minimum Rate Lifetime	Effective Date
1118-1950	72	1,000 hours	1/1/2012
788-1117	53	1,000 hours	1/1/2013
563-787	43	1,000 hours	1/1/2014
232-562	29	1,000 hours	1/1/2014

"modified spectrum" lamp.

Exemptions include rough vibration service, shatter-resistant, 3-way, black light, infrared, colored, bug, appliance, left-hand thread, marine and marine signal, mine service, plant light, reflector, sign service, silver bowl, showcase, traffic signal, G shape, T shape and AB, BA, CA, F, G16-1/2, G-25, G30, S and M-14 lamps. To prevent consumers shifting from standard incandescents to rough service, vibration service, 2601-3300 lumen general service and shatter-resistant lamps, sales of any of these lamp types will be monitored and if found to double above the increase modeled for a given year, the lamp type will become subject to rulemaking.

Manufacturers expect the market to shift to compact fluorescent after the Act goes into effect. LED lamps targeting replacement of incandescent lamps up to 60W are now being introduced by the major lamp manufacturers and may offer viable replacement, but at a price tag as high as $40+, adoption is not expected to be significant in the early adopter phase of the technology. However, self-ballasted compact fluorescents are likely to be resisted for the same reason they have stalled out in adoption by consumers: In some areas such as dimming and form factor, today's compact fluorescents are not an equivalent technology (although the technology is developing rapidly). For this reason, American consumers in 2012 may follow European consumers today by shifting to energy-saving halogen screw-in lamps that comply with the Act while offering 100 CRI, warm color temperature, compatibility with existing dimmers, and so on. Examples in the U.S. include Philips Energy Saver/Energy Advantage Halogená, Sylvania Halogen SuperSaver and models from GE that are expected to be introduced in the U.S. market in 2011 or 2012.

As a result, while the incandescent lamp's days may be numbered, those numbers are unlikely to be 2012, 2013 or 2014 as some expect. With halogen lamps on the market, self-ballasted compact fluorescents will have to fight for every socket as they are doing now. Recognizing these barriers to adoption, manufacturers are now working on compact fluorescents that are closer to the incandescent in performance, size and features. For example, GE's Energy Smart lamp offers consumers the same shape, size and aesthetic as an incandescent A19 or A21. Manufacturers are introducing more dimmable product, such as TCPs Super Dimmable lamps, which dim to 3 percent using a technology based on an electronic chip instead of a standard circuit board; the product starts quickly and dims linearly without color shift or flicker. And as of this writing, Sylvania recently announced a new 23W self-ballasted compact fluorescent lamp that dims to 20 percent using a dimmer knob at the base of the lamp, which can turn any household fixture into dimmable lighting, such as table lamps and floor lamps.

Energy Independence and Security Act of 2007 Eliminates Many Incandescent Reflector Lamps

Reflector lamp types are directional lamps—spots, floods, etc.—popular in recessed downlighting and track lighting applications, both residential and commercial. While general-service incandescent lamps have received the most attention in media coverage of the Energy Independence and Security Act of 2007, with provisions beginning to take effect in 2012, many popular incandescent reflector lamps were outlawed almost immediately, with further eliminations scheduled by subsequent DOE regulations (described later in this chapter).

With the Act's incandescent reflector lamp provisions taking effect, minimum efficacy standards established for incandescent reflector lamps by the Energy Policy Act of 1992 began to apply to a larger group of reflector lamps, creating a national standard matching stricter standards enacted by nine states since 2006.

Are CFLs Net Producers of Mercury?

After the Energy Independence and Security Act became law in late 2007, threatening an end to 40-100W general-service incandescent lamps, a consumer backlash began building against CFLs.

A common argument is that CFLs contain mercury, a toxic substance, whereas incandescents do not, countering the primary environmental benefit of using CFLs—significant reductions in greenhouse gas emissions at power plants.

The average CFL contains 4 mg of mercury, according to NEMA, after reductions following a voluntary commitment by its lamp manufacturing members in 2007 to cap mercury content at 5 mg for <25W CFLs sold into the residential market. Just enough to cover the tip of a ballpoint pen. Some products have reached as low as 1.5 mg.

During use and disposal, some CFLs are certain to break and emit mercury into the environment. According to the EPA, most of the mercury remains bound to the lamp if the bulb is broken, and the amount that escapes the lamp is still debated; emission estimates vary broadly from 1.2 to 6.8 percent. This means if a CFL containing 4 mg of mercury is not recycled and breaks, 0.05-0.27 mg may be emitted.

So incandescents are better than CFLs when it comes to mercury emissions, right? Not so fast. A problem with this argument is that coal-fired power plants produce about one-half of all electricity in the U.S. and, according to the U.S. Environmental Protection Agency (EPA), are the largest source of human-caused mercury emissions in the country—more than 50 tons in 2006. A portion of these emissions is airborne, oxidized and water-soluble; some ends up deposited in the U.S., while the rest enters the global cycle (more than half of mercury deposited in the U.S., for example, originates at Asian factories). Mercury released into the air is the main way it gets into water; eating contaminated fish is subsequently the main way humans become exposed.

Because incandescent lamps consume three to four times more energy than a CFL, they cause more atmospheric mercury emissions at power plants that burn coal. According to EPA's eGRID2006 (v.2.1), about 0.022 mg of mercury is emitted into the atmosphere per kWh of energy produced by coal-fired plants, or 0.012 mg per kWh averaged across all power generation in the U.S. A 75W incandescent operating over a period of 10,000 hours—the rated life of a competitive 18W CFL—will therefore generate an average 9.2 mg of atmospheric mercury emissions nationally, while the 18W CFL will generate 2.2 mg (plus up to another 0.27 mg due to breakage).

So CFLs produce less mercury nationally, right? Yes, while also slashing carbon emissions and providing substantial energy cost savings to their owners. But wait—what about states that don't burn coal for power? What about differing quality and mercury content of coal and mercury control technologies at power plants? And what about mercury emitted during lamp production? Looking at the question at this level of detail would require intensive modeling including a broad range of factors.

And a team of Yale researchers recently did just that. Key assumptions included 1) each CFL contains 5 mg of mercury, 2) 21 percent of spent lamps are recycled, 3) 25 percent of the mercury in CFLs that is not recycled becomes atmospheric emissions due to breakage during transit, 4) 3.5 percent of mercury entering landfills becomes atmospheric emissions, and 5) CFL manufacturing produces slightly higher emissions per lamp but less overall due to longer average life.

The researchers concluded that replacing a common 60W incandescent with a 15W CFL typically yields net reductions in atmospheric mercury emissions ranging from 0.01 to 20 mg in the U.S. over 10,000 hours—but nominally not in states that use very little coal to make electricity: Alaska, California, Oregon, Idaho, Vermont, New Hampshire, Maine and Rhode Island. In these states, switching to a CFL can actually cause a small net gain in atmospheric mercury emissions—up to about 1.2 mg.

So are CFLs net producers of atmospheric mercury emissions? Technically, the answer depends whether you are asking on a national, state or even utility level. As a national average and in most states, switching from incandescent to CFL results in significant reductions in mercury and other harmful emissions. And this is really a national issue, and anything else is just quibbling. After all, just as all generated power shares the same grid, we all share the same airflow.

Meanwhile, we can look forward to further reductions in mercury in the future as manufacturers continue to reduce the amount of mercury in their products, more attention is paid to lamp recycling, and the EPA's Clean Air Mercury Rule goes into final implementation in 2018, which will reduce mercury emissions from coal-fired plants by nearly 70 percent.

To see the Yale researchers' report, visit http://pubs.acs.org/cgi-bin/abstract.cgi/esthag/asap/abs/es800117h.html. To view detailed EPA-recommended procedures for cleaning up a broken CFL, visit www.energystar.gov.

In short, the Act covers incandescent and halogen reflector lamps greater than 2.25 inches in diameter, including R, PAR, BPAR, BR (BR30, BR40) and ER (ER30, ER40) lamps. After the effective date in June 2008, these lamp types were required to demonstrate a minimum efficacy as shown in **Table 4-5**, or no longer be manufactured in the United States. Any product manufactured prior to the effective date could continue to be sold until inventories are depleted.

Exemptions include <50W BR30, BR40, ER30, ER40; <45W R20;

Table 4-5. Energy Independence and Security Act of 2007 requirements for incandescent reflector lamps.

Wattage Range	Minimum Lumens/W
40-50W	10.5
51-66W	11.0
67-85W	12.5
86-115W	14.0
116-155W	14.5
156-205W	15.0

and 65W BR30, BR40 and ER40 lamps. The most popular lamps being eliminated included 50W and 75W R20 and 85W BR30 lamps, and the most popular exceptions are 65W BR30 and BR40 lamps. Color lamps and lamps designed for "vibration service" or "rough" applications are also exempt.

Energy Independence and Security Act of 2007
Eliminates Probe-start Metal Halide Luminaires

The Energy Independence and Security Act of 2007 contains significant provisions affecting the sale of metal halide luminaires. Starting in 2009, 150-500W metal halide luminaires must contain ballasts that operate at a certain level of efficiency, virtually eliminating probe-start lamps and ballasts from new luminaires. This provision of the Act essentially made a Federal standard of efficiency requirements already enacted in a number of states.

Specifically, starting January 1, 2009, new metal halide luminaires could no longer be manufactured or imported unless their ballast operated the lamp at a minimum efficiency level as shown in **Table 4-6**. Compliant luminaires contain a capital E printed in a circle on their packaging and ballast labels (similar to legislated fluorescent ballasts). Exceptions include luminaires with regulated lag ballasts, luminaires with electronic ballasts for operation at 480V, and 150W wet-location luminaires containing a ballast rated to operate at ambient temperatures above 50°C.

The Act covers manufacture and importation but not sale, so distributors are allowed to sell their inventories of non-compliant luminaires

Table 4-6. Minimum metal halide ballast efficiency levels mandated by EISA 2007. Efficiency is measured as Pout/Pin where Pout is the lamp wattage and Pin is operating wattage.

Ballast	Minimum ballast efficiency
Probe-start magnetic ballast	94%
Probe-start electronic ballast	90% if <250W
	92% if >250W
Pulse-start ballast	88%

until these are depleted (unless prohibited by state law). The law covers luminaires and not ballasts, so distributors will also be able to continue selling non-compliant ballasts to customers for spot replacement needs in existing installations.

As a result, probe-start magnetic ballasts for metal halide lamps up to 400W have been virtually eliminated from new luminaires. Because probe-start lamps require probe-start ballasts, this also eliminated 175-400W probe-start metal halide lamps from new luminaires. Demand is shifting to pulse-start. Note, however, that while many pulse-start magnetic ballasts comply with the Act, a significant number do not, so look for the compliance symbol on the ballast label and luminaire packaging. Meanwhile, most, if not all, pulse-start electronic ballasts comply, so it is expected that electronic ballasts, including dimming versions, will get a boost.

In recent years, pulse-start lighting systems have been making significant advances against traditional probe-start in higher wattages.

Pulse-start systems produce higher light output than traditional probe-start systems both initially and over time, operate more efficiently, produce whiter light, provide good lamp-to-lamp color consistency, and turn on and re-strike faster. In terms of energy efficiency, pulse-start metal systems can provide up to 25 percent energy cost savings in existing applications over probe-start, and up to 30 percent savings in capital and operating costs in new construction.

So the law is essentially accelerating the phase-out of an obsolete technology in favor of readily available, better-performing, more-efficient technologies.

Figure 4-1. Starting January 1, 2009, probe-start magnetic ballasts for operation of lamps up to 400W were virtually eliminated from new luminaires and, with them, most 175-400W probe-start metal halide lamps. Compliant alternatives include fluorescent and pulse-start metal halide. Photo courtesy of Universal Lighting Technologies.

California Title 20 Mandates Higher Efficiency for Metal Halide Luminaires

California new Title 20 standards, which went into effect January 1, 2010, created new energy efficiency standards for 150-500W metal halide luminaires used in indoor and outdoor applications. These luminaires may not be manufactured in the State of California unless they meet the new standards.

Indoor luminaires. First, no probe-start ballasts are allowed. Next, the luminaire must comply with the Energy Independence and Security Act of 2007, which imposes a minimum acceptable efficiency of 88 percent for the luminaire's pulse-start ballast.

But then Title 20 goes beyond EISA 2007, requiring one of four options for the luminaire.

1. Minimum ballast efficiency of 90 percent for 150-250W lamps and 92 percent for 251-500W lamps. Basically, by choosing a higher-efficiency ballast than that required by EISA 2007, the Title 20 requirement can be satisfied.

2. A ballast with an efficiency of 88 percent or greater AND an integral occupancy sensor with a default setting to automatically reduce lamp power through dimming by at least 40 percent within 30 minutes or less of an area being vacated.

3. A ballast with an efficiency of 88 percent or greater AND an integral photocontrol to automatically reduce lamp power through dimming by at least 40 percent in response to daylight contribution to light levels.

4. A ballast with an efficiency of 88 percent or greater AND a relamping rated wattage (stated on a permanent, pre-printed factory-installed luminaire label) with only one of these four bins: a) 150-160W, b) 200-215W, c) 290-335W or d) 336-500W (if the luminaire is able to operate 336-500W lamps, it must be prepackaged and sold together with at least one lamp per socket with a minimum lamp mean efficacy of 80 lumens/W.

Basically, the ballast must be even more efficient than EISA 2007 or use lighting controls. The intent appears to be to push end-users towards use of 150-500W electronic ballasts, which currently represent 2 percent of total HID ballast shipments, according to 2009 NEMA data. Alternately, users can stick with magnetic ballasts that offer a compliant level of efficiency or control capability. Magnetic ballasts would be most desirable for applications where electronic ballasts are not yet available or where the alternative has a form factor requiring modification of the luminaire.

In addition, electronic ballasts are still proving themselves in extreme environments in which magnetic HID ballasts have already proven themselves.

Of course, there are exceptions, which negate the minimum ballast efficiency requirements for certain metal halide lighting systems if they meet any of the following conditions:

1. Luminaires that use regulated lag ballasts.
2. Luminaires that use electronic ballasts that operate at 480V.
3. Luminaires that a) are rated for use only with 150W lamps, b) are rated for use in wet locations [as specified by NEC 2002, Section 410.4(A)], AND c) contain a ballast that is rated to operate at ambient air temperatures above 50C, as specified by UL 1029-2001.

Outdoor luminaires. Same as the above, but outdoor luminaires that may comply with Option #3 must comply with at least one of the other options.

Future requirement. Starting January 1, 2015, indoor 150-500W metal halide luminaires must comply with Option #4 (wattage bin ranges) in addition to at least one of the other compliance options.

Department of Energy Rules to Impact
General-service Fluorescent Lamps

In 2009, DOE announced new efficiency standards for commercial general-service fluorescent and incandescent reflector lamps. The result for each lamp type is that the least-efficient and lowest-cost products will be removed from the market when the rules become effective July 14, 2012.

DOE rules announced in 2009 expand upon efficiency rules established by the Energy Policy Act of 1992 by strengthening standards for covered lamp types while also covering 8-ft. T8 lamps, 4-ft. T5 lamps and more wattages of 4-ft. T8 and T12 lamps.

The net result is a majority of 4-ft. linear and 2-ft. U-shaped T12 lamps, many 8-ft. T12 and T12HO, and some lower-color-rendering 4-ft. T8 lamps will be eliminated when the rules become effective July 14, 2012, including:

*   Most 4-ft. F40 and F34T12 linear lamps
*   All 2-ft. FB40 and FB34T12 U-shaped lamps

- All 75W F96T12 and 110W T96T12HO lamps
- Most 60W F96T12/ES and 95W F96T12/ES/HO lamps
- All 4-ft. T8 basic-grade 700/SP series lamps rated at 2,800 lumens
- Some 8-ft. T8 Slimline single-pin 700/SP series and 8-ft. T8 HO RDC-base lamps

Current exemptions will continue to be recognized, including lamps with a CRI rating of 87 or higher, lamps designed to operate in cold temperatures, ultraviolet lamps and certain other specialty lamps.

What this means for building owners is that fewer lamp choices will be available to purchase as the least-efficient and lowest-cost products are removed from the market, with the new baseline being lamps that are more efficient and often offer greater color rendering ability and/or longer service life. At least one lamp manufacturer is expected to offer a compliant 4-ft., 34W, T12 lamp, but it will be more expensive than today's lamps due to the use of rare earth phosphors. Distributors will not be prohibited from selling non-compliant lamps after the effective date, so they will be able to continue selling them until their inventories are exhausted.

Department of Energy Rules to Impact
Incandescent Reflector Lamps

In 2009, DOE announced new efficiency standards for incandescent reflector lamps that become effective July 14, 2012. The new DOE rules expand upon efficiency rules established by the Energy Policy Act of 1992 and the Energy Independence and Security Act of 2007 (see above) by strengthening standards covering incandescent reflector lamps.

The result is a majority of incandescent and halogen reflector lamps will be eliminated, including the market's least-efficient and lowest-cost products. The manufacturing community will have to produce alternatives. Most affected lamps are 40-205W R, PAR, BR, ER and BPAR lamps with a diameter larger than 2.5 in. This includes 130V lamps operated on 120V with the intention of doubling lamp life with a modest reduction in light output.

Existing exemptions included in the Energy Independence and Security Act of 2007 are expected to remain intact until July 2013, per pending energy legislation. These lamps, common in residential and some commercial applications, include 50W and lower BR30, BR40, ER30 and ER40; 45W and lower R20; and 65W BR30, BR40 and ER40 lamps.

Building owners already adapting to the Energy Independence and

Security Act of 2007's incandescent reflector lamp requirements will essentially have fewer halogen options left that comply. If the same qualities as halogen are desired—such as dimmability, lighting quality and intensity—they can substitute halogen infrared reflector lamps that pass the new standards (although not all will). In this type of lamp, the IR coating redirects wasted heat produced during light emission back to the filament—raising its temperature and enabling the lamp to produce more light output for the same input watts—boosting efficacy by 20-30 percent, according to one manufacturer. Alternately, low-voltage halogen can be recommended.

Other options include compact fluorescent, self-ballasted metal halide and LED lamps, but these currently do not provide performance equivalent to halogen.

As with the fluorescent lamp rules, the incandescent reflector lamp rules will result in a new baseline that is more efficient and more expensive than today's least-efficient and lowest-cost lamps. Distributors will be able to continue selling non-compliant product after the effective date until inventories are exhausted. Unlike fluorescent lamps, however, there are few alternatives in the same lamp family that exist today, which means manufacturers will be developing new products based on demand over the next two and a half years, and providing guidance to contractors on substitutions. By then, it may also be possible that there will be other technologies that can provide equivalent performance for much higher efficiency.

NEW YORK CITY: LARGEST BUILDINGS
MUST UPGRADE THEIR LIGHTING

New York City's dense urban landscape is populated with more than one million buildings that annually consume $15 billion in energy and generate 75 percent of the City's carbon emissions. The 22,000 largest buildings, concentrated largely in Manhattan, account for roughly 45 percent of total floorspace and energy consumption. Lighting is a major energy user, responsible for nearly 18 percent of energy consumption and resulting carbon emissions.

To reduce energy costs and carbon emissions, the New York City Council enacted major energy legislation on December 9, 2009 that includes an ambitious law requiring large commercial buildings to upgrade their lighting systems. While this legislation has the immediate effect of creating instant opportunities for electrical contractors doing business in

Figure 4-2. With some notable exceptions, a number of incandescent reflector lamps popular in track and downlighting were phased out beginning in June 2008, per new efficiency standards in the Energy Independence and Security Act of 2007. Subsequently, DOE announced strengthened standards covering incandescent reflector lamps that will go into effect July 14, 2012. The net result is that with some exceptions, incandescent and halogen reflector lamps will no longer be manufactured unless they meet stringent efficiency standards. Options include high-efficiency infrared-coated halogen lamps, compact fluorescent, self-ballasted metal halide and LED lamps. Photo courtesy of OSRAM SYLVANIA, Inc.

the City, it may have a more long-range and far-reaching effect if the policy proves successful in getting existing buildings to improve their energy efficiency.

Int. No. 973-A, which took effect immediately, requires large commercial buildings to upgrade their lighting by January 1, 2025. Specifically, the law applies to buildings larger than 50,000 sq.ft., buildings combining with other buildings on the same tax lot to exceed 100,000 sq.ft. in total, and buildings held in the condominium form of ownership governed by the same board of managers and together exceed 100,000 sq.ft. in total. Another provision in this law requires these buildings to submeter tenant spaces larger than 10,000 sq.ft. and provide this information to the tenants, including monthly statements of electricity consumption and costs.

The law defines a lighting upgrade as meeting the minimum requirements of the New York City Energy Conservation Code. Exceptions include residential living spaces; spaces serving these living spaces such as laundry rooms, boiler rooms and hallways, stairways and corridors used for egress; emergency or security areas; assembly spaces in houses of worship; and lighting that meets code installed on or after July 1, 2010. The code itself has its own exceptions.

The New York City Energy Conservation Code is based on the New York State energy code, with amendments making it more stringent. The 2007 state code is in turn based on the 2003 version of the International Energy Conservation Code (IECC) model energy code, with amendments, while referencing ASHRAE 90.1-2004 as an alternative standard. The lighting section of the Code includes mandatory and prescriptive requirements for lighting controls (interior lighting controls, light level reduction controls and automatic lighting shutoff), tandem wiring, exit signs, interior lighting power caps and exterior lighting.

To demonstrate compliance, the owner must file a report with the New York City Department of Buildings, prepared by a registered design professional or a licensed master or special electrician, certifying that the lighting upgrade has been completed and that the work is in compliance with the technical standards of the New York City electrical code.

Speaking of the City's energy code, Int. 564-A, enacted at the same time, refreshed the New York City Energy Code, closing the loophole enabling buildings to perpetuate non-compliant building systems if the owner performs a renovation on less than half of a given building system. Out of more than 40 states based on the IECC, only New York State exempted minor renovations. Starting July 1, 2010, all renovations requiring a building permit, even small renovations, must demonstrate compliance with the City's energy code.

Int. 973-A, however, does not explicitly spell out what version of the code must be complied based on when the lighting upgrade is completed. If the upgrade occurs in 2024, does the owner have to comply with today's energy code or whatever version of the code is in effect in 2024? As it is written, the language suggests whatever version is in effect at the time. This means building owners interested in managing risk might find it prudent to upgrade now rather than putting it off to the last minute, or else be forced to upgrade to a much stricter code than what is now in effect.

And that is not all that is required. Int. No. 476-A requires large building owners to make an annual benchmark analysis of energy consumption to enable owners, tenants and potential tenants to compare energy consumption of different buildings. And Int. No. 967 requires large private buildings to conduct energy audits once every 10 years and implement energy-efficient maintenance practices; all City-owned buildings over 10,000 sq.ft., meanwhile, will be required to conduct audits and complete energy-efficient retrofits that pay for themselves within seven years.

Critics of the legislation say building owners have too long to upgrade their lighting and that it lacks a mechanism forcing them to do it. Proponents of the legislation welcome what is arguably the country's most ambitious initiative to increase the energy efficiency of existing buildings, where energy efficiency measures can have the biggest impact.

COMMERCIAL BUILDINGS TAX DEDUCTION EXTENDED TO 2013

The Energy Policy Act of 2005 created the Commercial Buildings Deduction (CBD) an incentive consisting of an accelerated tax deduction rewarding investment in energy-efficient interior lighting, HVAC/hot water systems and building envelope.

Initially set to expire December 31, 2007, the CBD was extended to December 31, 2008 by Congress and, with the passage of Stimulus legislation in 2009, was extended another five years to December 31, 2013.

The CBD includes a tax deduction for reducing energy and power costs to at least 50 percent less than a building satisfying the ASHRAE 90.1-2001 standard. A partial deduction is also available for investments in interior lighting only. And the Interim Lighting Rule, which was supposed to be in effect only until the partial deduction rules were written, is still in effect. Of the three CBD options, the Interim Lighting Rule is the most straightforward because it deals only with lighting and can be calculated using a standard spreadsheet program, with no building energy modeling being required.

The Interim Lighting Rule enables owners of commercial buildings to deduct the full cost of new interior lighting, capped at $0.30-$0.60/ sq.ft. on a sliding scale, if the new lighting achieves a lighting power density (W/sq.ft.) that is 25-40 percent lower than the maximum values published in ASHRAE 90.1-2001's Table 9.3.1.1 or Table 9.3.1.2. The exception is warehouses: The lighting system must reduce power density by at least 50 percent to earn a deduction of up to $0.60/sq.ft.

Qualifying building types are listed in Table 9.3.1.1, although IRS Notice 2008-40 adds nonresidential unconditioned garage spaces to building types covered by the Rule. Houses of worship, meanwhile, do not qualify because religious organizations are tax-exempt and their buildings are not owned by the public.

In a new building or major renovation that triggers the application of the local commercial building energy code, the project may already come close to satisfying the requirements of the CBD. The 2004 and 2007 versions of 90.1 contain lighting power density caps that are generally 13-50 percent lower than the 2001 version. For example, maximum allowable power density for office buildings is 23 percent lower. Retail 21 percent lower, hospitals 25 percent lower, manufacturing 41 percent lower, schools and universities 20 percent lower, and warehouses 33 percent lower. This means that simply by designing the lighting system to the requirements of an energy code based on ASHRAE 90.1-2004 or 2007, the design may already be near the zone of qualification.

Besides achieving a reduction in power density, three other conditions must be met. First, if the project triggers the controls provisions of ASHRAE 90.1-2001, these provisions must be met. Second, bi-level switching must be installed in all occupancies except hotel and motel guest rooms, store rooms, restrooms and public lobbies. And third, the application must meet the minimum requirements for calculated light levels as published in the ninth edition of the IES *Lighting Handbook*.

What is "bi-level switching?" NEMA has recognized a definition of bi-level switching as manual and/or automatic control that provides at least two levels of lighting power in a space—not including OFF—delivered by permanently installed lighting. This could involve lighting layers, A/B switching and step- or continuous dimming, with the control input being a switch, timer, photocontrol, occupancy sensor, lighting automation panel or some other device (for more on controls, see the author's *Lighting Controls Handbook*).

If the building is government-owned (does not pay taxes), the designer ("person that creates the technical specifications for installation of energy-efficient commercial building property") can claim the deduction, according to IRS Notice 2008-40. If more than one designer is involved, the owner may allocate the deduction to the designer who was primarily responsible for the design or, at the owner's discretion, among the designers.

For a building owner to claim the CBD, the project must be certified by a qualified individual—a contractor or engineer properly licensed as such in the jurisdiction where the building is located. This individual, who cannot be an employee of the building owner, must demonstrate in writing to the owner that they have the qualifications to do the certification.

According to Notice 2008-40, the certification of an Interim Lighting Rule project must include a statement that the interior lighting achieves a suitable reduction in power density and satisfies the mandatory requirements for lighting controls and calculated light levels. It must also include a statement that a field inspection was performed by a qualified individual after the lighting was installed, in accordance with the procedures contained in *Energy Savings Modeling and Inspection Guidelines for Commercial Building Federal Tax Deduction* published in February 2007, and that the expected energy savings are being realized. Finally, it must include a list of energy-efficient lighting components, an explanation of these features, and projected lighting power density.

Note that although IRS Notice 2006-52 says the certification must include a statement that qualified computer software was used to calculate energy and power consumption and costs, this is not needed to demonstrate compliance with the Interim Lighting Rule. Instead, a spreadsheet or basic lighting analysis software can be used.

It took a while, but the stars finally aligned for lighting opportunities available under the Energy Policy Act of 2005's Commercial Buildings Deduction. The Interim Lighting Rule can now be implemented with all required pieces in place and an expiration in 2013.

For more information, visit www.lightingtaxdeduction.org.

ENERGY CODES

Some designers consider them a necessary evil, others just another set of requirements to obtain a building permit. Energy codes are designed to set minimum standards for design and construction and can significantly reduce building system life-cycle costs.

U.S. energy codes address lighting prescriptively by setting lighting power density (LPD) limits on lighting for whole buildings. Setting LPD limits for whole buildings is important because energy-efficient lighting can be inefficient as a whole if installed in high densities in a building. Most codes now also mandate automatic lighting shutoff controls as well as controls for enclosed spaces.

Before 1992, states in the United States enacted energy codes on a voluntary basis, some developing their own codes while others adopted model codes and energy standards. Currently, the two major model codes and energy standards are:

- ASHRAE/IES Standard 90.1 *Energy-Efficient Design of New Buildings Except Low-Rise Residential Buildings*, developed jointly by American Society of Heating, Refrigerating and Air-Conditioning Engineers (ASHRAE) and the Illuminating Engineering Society (IES), was first published in 1975. The Standard was subsequently updated in 1980, 1989, 1999, 2001, 2004 and 2007; after 2001, the intention is to update the Standard every three years. At the time of writing, the 2010 version was on schedule for publication in December 2010.

- The International Energy Conservation Code (IECC), developed by the International Code Council (ICC), a membership association dedicated to building safety and fire prevention, is a model energy code that covers lighting in addition to other energy-using building systems. In 1998, the ICC released the IECC, followed by a 2000 version (with 2001 supplement), 2003 version (with 2004 supplement), 2006 version (with 2007 supplement), and 2009 version. The IECC references Standard 90.1 as an alternative path of compliance.

The Energy Policy Act of 1992 authorized DOE to establish a national energy standard for all states. Since 2004, Standard 90.1-1999, developed jointly between the American Society of Heating, Refrigerating and Air-Conditioning Engineers (ASHRAE) and IES, is the national energy standard. All states must have a commercial energy code at least as stringent as this. At the end of 2008, DOE recognized ASHRAE 90.1-2004 as the new national energy standard, which goes into effect December 30, 2010.

Despite the existence of a national energy standard, the country is a patchwork of energy codes. Some states have adopted Standard 90.1-1999, while others have adopted the 2001, 2004 or 2007 versions (see **Table 4-7**). Many have adopted the IECC. Often, these codes are adopted with amendments that individualize their codes. Other states, such as California, Florida, Oregon and Washington, have developed their own codes. And a significant number of states have not complied with the DOE mandate. As of the time of writing, the most common codes are IECC 2003 and Standard 90.1-2004. Although there has been some convergence towards consistent requirements (IECC references Standard 90.1 as an alternative compliance standard, for example), these differences can be confusing for

Table 4-7. Status of commercial state energy codes as of May 2010, adapted from www.energycodes.gov. For the most up-to-date list, visit www.energycodes.gov.

State	Status	State	Status
Alabama	No statewide code	Montana	2009 IECC with reference to ASHRAE 90.1-2007
Alaska	2006 IECC (public buildings only)	Nebraska	2003 IECC with reference to ASHRAE 90.1-2004
Arkansas	ASHRAE 90.1-2004 (commercial), ASHRAE 90.1-2007 (public buildings)	Nevada	2006 IECC with reference to ASHRAE 90.1-2004
Arizona	Significant adoption in local jurisdictions	New Hampshire	2009 IECC with reference to ASHRAE 90.1-2007
California	State-specific code: Title 24-2008 that exceeds ASHRAE 90.1-2007	New Jersey	ASHRAE 90.1-2004 with amendments
Colorado	Voluntary state provisions based on 2003 IECC with reference to ASHRAE 90.1-2001	New Mexico	2006 IECC
Connecticut	2006 IECC with reference to ASHRAE 90.1-2004	New York	2003 IECC with amendments
Delaware	2009 IECC	North Carolina	State code modeled on 2006 IECC with amendments including ASHRAE 90.1-2004
Florida	State-specific code more stringent than ASHRAE 90.1-2007	North Dakota	No statewide code
Georgia	IECC 2006 with reference to ASHRAE 90.1-2004, with amendments	Ohio	2006 IECC with reference to ASHRAE 90.1-2004
Hawaii	Mix of 2006 and 2009 IECC and ASHRAE 90.1-1999 by county	Oklahoma	2006 International Building Code with default to 2006 IECC
Idaho	2006 IECC with reference to ASHRAE 90.1-2004	Oregon	State-specific code meets or exceeds ASHRAE 90.1-2004
Illinois	2009 IECC with reference to ASHRAE 90.1-2007	Pennsylvania	2009 IECC with reference to ASHRAE 90.1-2007
Indiana	ASHRAE 90.1-2007	Rhode Island	2006 IECC with reference to ASHRAE 90.1-2004
Iowa	2009 IECC with reference to ASHRAE 90.1-2007	South Carolina	2006 IECC
Kansas	2006 IECC	South Dakota	No statewide code
Kentucky	2006 IECC	Tennessee	ASHRAE 90A/90B (very old standard)
Louisiana	ASHRAE 90.1-2004	Texas	2000 IECC with 2001 supplement
Maine	2009 IECC with reference to ASHARAE 90.1-2007	Utah	2006 IECC with reference to ASHRAE 90.1-2004
Maryland	2009 IECC with reference to ASHARAE 90.1-2007	Vermont	Code based on 2004 IECC with amendments to include ASHRAE 90.1-2004
Massachusetts	2009 IECC with amendments	Virginia	2006 IECC with reference to ASHRAE 90.1-2004
Michigan	ASHRAE 90.1-1999	Washington	State-specific code more stringent than ASHRAE 90.1-2004
Minnesota	ASHRAE 90.1-2004	West Virginia	2003 IECC with amendments
Mississippi	ASHRAE 90.1-1975	Wisconsin	2006 IECC with amendments
Missouri	Significant adoption in local jurisdictions	Wyoming	No statewide code

designers working in multiple jurisdictions.

With various rules and exceptions, IECC and 90.1 present the lighting requirements in **Table 4-8**.

Effect of LEED on State Energy Code Requirements

The USGBC's Leadership in Energy and Environmental Design (LEED) green building rating system is referenced in dozens of ordinances, executive orders, policies and incentives in 34 states, 138 cities and 14 Federal agencies and departments as of the time of writing. Since LEED is based on meeting and exceeding ASHRAE 90.1-2007, this may effectively implement another energy code in the jurisdiction—if, for example, a given state has a commercial building code based on 2003 IECC but, with LEED Silver required for all state buildings, also has a code for state building construction based on ASHRAE 90.1-2007.

In New Jersey, for example, all new state construction 15,000 sq.ft. and larger must earn LEED Silver certification or equivalent, according to the USGBC. In Atlanta, all City-funded projects larger than 5,000 sq.ft. are required to achieve LEED Silver, while Los Angeles is requiring all City buildings larger than 7,500 sq.ft. or built before 1978 to be retrofitted to LEED for Existing Buildings Silver certification. At NASA, new construction and major renovation projects at the Agency's facilities are required to achieve LEED Silver and strive for Gold.

Most of these policies affect public construction. Private construction, when addressed, is usually encouraged to achieve LEED certification or an equivalent using incentives, such as tax credits. Some jurisdictions are beginning to place similar requirements on private construction, however—such as Albany, CA, which requires LEED certification for all private commercial projects 10,000 sq.ft. and larger, and Boston, which requires public and private development projects over 50,000 sq.ft. to earn either LEED certification or approval based on another method, according to USGBC.

As green building construction standards (discussed later in this chapter) become adopted as codes, the situation of what code applies and to what project should crystallize.

Energy Codes and Existing Buildings

Model energy codes and standards cover new construction and renovations. As of the time of writing, many types of retrofits, such as replacing lamps and ballasts, are not directly impacted by energy codes,

Table 4-8. Summary of major lighting requirements in IECC and 90.1.

	IECC			ASHRAE/IESNA			
	2003	2006	2009	1999	2001	2004	2007
Automatic Lighting Shutoff							
Scheduling OR	x	x	x	x	x	x	x
Occupancy sensors OR	x	x	x	x	x	x	x
Another automatic method	x	x	x	x	x	x	x
Explicitly states that furniture-mounted supplemental task lighting load is not counted in interior lighting power if controlled by automatic shutoff such as occupancy sensor			x				x
Space Controls							
Manual	x	x	x	x	x	x	x
Light reduction control (dimming or switching, unless an occupancy sensor is installed)	x	x	x				
Employee lunch/break room, higher education classroom and conference/meeting room lighting must be operated by either multi-scene control or occupancy sensor						x	x
Daylighting Harvesting Control							
Independently controlled general lighting in "daylighting zones" (as defined by code)			x				
Guest Rooms/Sleeping Units (Hotels, Motels, etc.)							
Must have control at room entrances to turn off permanently installed and switched receptacle lighting	x	x	x	x	x	x	x
Display/Accent, Non-Visual (Such as for Plant Growth), Demonstration Lighting							
Separate, independent control required				x	x	x	x
Task Lighting							
Integral control device or readily accessible wall-mounted device				x	x	x	x
Exterior Lighting Control							
Photosensor OR	x	x	x	x	x	x	x
Astronomical time switch OR			x			x	x
Combination of photosensor and astronomical time switch			x				x
Tandem Wiring							
Tandem wiring requirements	x	x	x	x	x	x	x
Exit Signs							
Minimum source efficacy of 25 LPW				x	x		
No more than 5W per face	x	x	x			x	x

(Cont'd)

Table 4-8 (*Cont'd*). Summary of major lighting requirements in IECC and 90.1.

Exterior Building Grounds Lighting (other than low-voltage landscape luminaires)							
Luminaire >100W must have source efficacy of 45 LPW	x						
Luminaire >100W must have source efficacy of 60 LPW OR		x	x	x	x	x	x
Luminaire>100W must be controlled by motion sensor		x	x	x	x	x	x
Interior Lighting Power							
Building method OR	x	x		x	x	x	x
Space by space method OR				x	x	x	x
Performance method	x	x	x	x	x	x	x
Exterior Lighting Power: Prescriptive LPD allowances							
Exterior lighting LPD allowances		x	x	x	x	x	x
Electrical Energy Consumption							
If building contains individual dwelling units, each unit must be separately metered	x	x	x				

although some jurisdictions may require a building permit for some lighting upgrades. With publication of the ASHRAE 90.1-2010 standard in the fall of 2010, this may begin to change, as 90-2010 is expected to include lamp and ballast retrofits in some manner. Below is a discussion of IECC and 90.1 language that applies to existing buildings. Any interpretations are solely the author's, and are provided here solely for educational purposes; of course, the building official or other authority having jurisdiction (AHJ) has the final say.

Let's begin with IECC 2006, which states:

"Except as specified in this chapter, this code shall not be used to require the removal, alteration or abandonment of, nor prevent the continued use and maintenance of, an existing building or building system lawfully in existence at the time of adoption of this code."

In other words, simply, if the building already exists, it exists, and the code by itself does cannot compel its owner to do anything to it. The model code, however, goes on to state:

"Additions, alterations, renovations or repairs to an existing building, building system or portion thereof shall conform to the provisions of this code as they relate to new construction without requiring the unaltered portion(s) of the existing building or

building system to comply with this code. Additions, alterations, renovations or repairs shall not create an unsafe or hazardous condition or overload existing building systems."

So if a portion of the lighting system is altered, that portion must comply with the code and the alteration and subsequent code compliance cannot negatively impact safety, which might be interpreted to include reductions in light levels below IES-recommended levels. Unfortunately, the language—if you touch it, you must comply with code—is somewhat vague about whether lamp and ballast retrofits are covered. Some might argue that replacing lamps and ballasts is maintenance and not an alteration or repair, while others might argue that the lamp and ballast does not constitute a complete system. For these reasons, lamp and ballast retrofits are often considered not covered by IECC. Again, the AHJ has the final say. The code also stipulates:

"Buildings undergoing a change in occupancy that would result in an increase in demand for either fossil fuel or electrical energy shall comply with this code."

ASHRAE 90.1's language is more specific about lighting alterations. Section 9.1.2 of ASHRAE 90.1-2004 states:

"The replacement of lighting *systems* in any building space shall comply with the lighting power density requirements of Section 9 applicable to that space. New lighting *systems* shall comply with the applicable lighting power density requirements of Section 9."

At first, this sounds like it would include lamp and ballast retrofits. However, the 90.1 requirement defines "system" as a "combination of equipment and auxiliary devices … by which energy is transformed so it performs a specific function such as … lighting." While "combination of equipment and auxiliary devices" may sound somewhat open to interpretation, the author understands the intent is a system consists of luminaires, meaning system replacement involves luminaire replacement. There is also a major exception to this requirement:

"Alterations that replace less than 50 percent of the luminaires in a space need not comply with these requirements provided that such alterations do not increase the installed interior lighting power."

So the standard does not appear to apply to lamp and ballast retro-fits nor does it apply to any spaces in projects in which less than half of the luminaires are being replaced. Regarding lighting controls:

"Any new control devices as a direct replacement of existing control devices shall comply with the specific requirements of [an applicable section of the standard relating to controls]."

Lighting controls must comply with code if they replace existing controls. But the same exception applies, so lighting controls must comply with code only if 1) they replace existing controls and 2) the alteration also involves 50 percent or more of the luminaires being replaced in whatever portion of the lighting system is being altered.

ASHRAE 90.1-2010, still in development at the time of writing, is expected to cover both indoor and outdoor lamp/ballast retrofits and all luminaire replacement, with the lighting in the space required to comply with the LPD and automatic shutoff requirements of the Standard. The exception would be alterations that involve less than 10 percent of the connected lighting load, as long as those alterations do not increase the installed LPD.

What's New for Lighting in IECC 2009?

IECC 2009 contains a number of significant changes related specifically to lighting, including:

- Creation of daylight control zones
- More exemptions to interior lighting wattage
- Revised additional retail display allowances
- Exterior power allowances now based on system of outdoor lighting zones

Daylight control zones. Daylight can increase light levels in indoor spaces, which creates opportunities for a control system to reduce electric light output and thereby save energy. California's Title 24 requires separate zoning and control of lighting in daylight zones, IECC 2009 followed suit, and ASHRAE 90.1-2010 is likely to join them.

IECC 2009 requires lighting in zones adjacent to windows and skylights to be separately circuited and controlled from the general lighting. The size of the zone is defined by simple formulas.

While zoning is mandated, a control method is not. This means the designer can choose manual or automatic switches or dimming, whatever fits the application needs best.

Outdoor lighting control. All outdoor lighting, not specifically exempted by IECC 2009, must be turned off automatically either when there is sufficient daylight available or the lights are no longer required to be ON during the night.

The previous version of IECC says that non-dusk-to-dawn outdoor lighting must be controlled by an astronomical time switch. IECC 2009 says these luminaires can be controlled by either a time switch or a combination of a time switch and photosensor.

Allowing luminaires with combined control provides a little more flexibility by allowing the user to decide which is best for the application.

Furniture-mounted task lighting. IECC describes how to add up the total connected indoor lighting power and divide it by square footage, which must then be at or below the maximum allowed for each building type. IECC 2009 creates several new exceptions. These exceptions are valuable because if a lighting element does not have to be counted, the designer has that much more flexibility in designing the lighting system.

One important new exception is furniture-mounted supplemental task lighting that is controlled by automatic shutoff. If the task lighting is automatically turned off when it is not used by 1) an occupancy sensor, 2) schedule-based control system or 3) other method, then it does not have to be counted as part of the interior lighting power.

Line-voltage track lighting. In previous versions of IECC, line-voltage track lighting must be counted as contributing at least 30W/linear foot of track to the installed lighting power. This penalizes track installations with an actual wattage below 30W/linear foot.

Manufacturers began offering devices that limit the current available for the track section to a load closer to the actual lighting design. IECC 2009 recognizes these methods by stating the input watts for line-voltage track luminaires can be 30W/linear foot or the wattage limit of either 1) the system's circuit breaker or 2) another permanent current-limiting device on the system.

Products are available from a number of track lighting and control manufacturers. Solutions include subpanels with current-limiting breakers and current-limiter track connectors with a resettable breaker that trips if the connected load exceeds the current limit.

Retail display lighting. IECC provides additional lighting power allowances for retail spaces. IECC 2009 says separately controlled, non-general lighting used to highlight merchandise in sales areas can claim additional lighting power limited by the values in **Table 4-9**. These changes were made to increase application clarity and simplicity while reducing the potential for overlighting.

The way the additional lighting power allowances is set up is similar to ASHRAE 90.1-2007, but the actual values are significantly lower due to an emphasis on more-efficient options such as ceramic metal halide lighting.

Outdoor building lighting power. IECC 2009 includes a dramatically different methodology for calculating outdoor building lighting power based on outdoor lighting zones.

Total outdoor lighting power is defined as the base site allowance plus permitted individual allowances based on lighting zone. The base site allowances are shown in Table 505.6.2 and should be familiar to designers who have worked with previous versions of IECC. The individual allowances are based on whether the application is in Lighting Zone 1, 2, 3 or 4, as shown in **Table 4-10**.

Table 4-9. Additional power allowances for retail display lighting.

Products sold in sales area	Classification	Additional Lighting Power Allowance = 1000W + ...
All products not listed below	Retail Sales Area 1	(Floor Area x 0.6W/sq.ft.)
Vehicles, sporting goods, small electronics	Retail Sales Area 2	(Floor Area x 0.6W/sq.ft.)
Furniture, clothing, cosmetics, artwork	Retail Sales Area 3	(Floor Area x 1.4W/sq.ft.)
Jewelry, crystal, china	Retail Sales Area 4	(Floor Area x 2.5W/sq.ft.)

Table 4-10. Outdoor lighting zones.

Lighting Zone	Description
1	Developed areas of national parks, state parks, forest land and rural areas
2	Areas predominantly consisting of residential zoning, neighborhood business districts, light industry with limited nighttime use, and residential mixed-use areas
3	All other areas
4	High-activity commercial districts in major metropolitan areas as designated by the local land use planning authority

Zone 1 includes rural areas that would have lower light level needs. Most applications fall into Zone 2 or 3, which includes residential, neighborhood business and mixed-use areas. Zone 4 includes high-activity commercial districts where light level needs, and therefore power allowances, tend to be higher.

In the Spotlight: IECC 2009's Reference of 90.1 as an Alternative Standard

Previous versions of IECC allowed various construction disciplines—lighting, mechanical, envelope—to be able to comply with either the applicable version of IECC or designated version of ASHRAE 90.1. IECC 2009 changed that with Section 501.2 (Application):

"The commercial building project shall comply with the requirements in Sections 502 (Building envelope requirements, 503 (Building mechanical systems), 504 (Service water heating) and 505 (Electrical power and lighting systems) in its entirety. As an alternative the commercial building project shall comply with the requirements of ASHRAE/IESNA 90.1 in its entirety."

What this means is that all design disciplines have to comply with one code. Different disciplines have found more appropriate application in one code or the other, but now they are forced to choose. While easier to implement from the perspective of a code official, this restriction could limit design flexibility. Specifically, if the IECC path is chosen, lighting designers will not have the Space by Space Method as an option for compliance, as it is not included in IECC 2009. As a result, whoever designs the lighting system will need to negotiate with the project's architect and mechanical designer to determine the best path for code compliance of the overall project, which may in some cases involve some concessions in design flexibility.

What's New for Lighting in ASHRAE 90.1-2004?

As of the time of writing, ASHRAE 90.1-1999 is the national energy standard of record for state code adoption requirements in the United States. Starting July 2004, all states were required to implement a commercial energy code that is at least as stringent as 90.1-1999.

On December 30, 2008, DOE determined that switching to 90.1-2004 would generate site energy savings of about 12 percent, and recognized 90.1-2004 as the new national energy standard effective two years later, on

December 30, 2010.

Besides minor text clarifications and additional exceptions, 90.1-2004 includes three major changes compared to ASHRAE 90.1-1999. First, LPD values were completely revised (more restrictive). Second, outdoor lighting requirements were expanded to cover effectively any outdoor lighting application. Third, occupancy sensors are required in a number of spaces.

Automatic lighting shutoff. In both ASHRAE 90.1-1999 and 2004, interior lighting must be shut off when it is not being used in buildings larger than 5,000 sq.ft. The 1999 standard lists "by occupant intervention" as a method of control. This was an error in description that was corrected for the 2001 and beyond versions by replacing it with "a signal from another control or alarm system that indicates the area is unoccupied." For example, the last person to leave the building could activate the security system, which would signal the lighting to turn OFF.

More, 90.1-1999 recognizes only one exception, which is lighting intended for 24-hour operation. The 2004 standard adds "lighting in spaces where patient care is rendered" and "spaces where an automatic shutoff would endanger the safety or security of the room or building occupant(s)."

Space control. The 2004 standard changes the "space control" requirements by identifying three applications where an occupancy sensor is a required control to turn the lights OFF within 30 minutes of the occupant leaving the space:

• classrooms (but not including shop classrooms, laboratory classrooms, and preschool through 12th grade classrooms);
• conference/meeting rooms; and
• employee lunch and break rooms.

ASHRAE 90.1-2004 adds, "These spaces are not required to be connected to other automatic lighting shutoff controls."

Exit signs. The 1999 standard mandates a light source efficacy of at least 35 lumens/W for exit signs larger than 20W. ASHRAE 90.1-2004 simplifies and further tightens this requirement to read, "Internally illuminated exit signs shall not exceed 5 watts per face." Basically, this means LED.

Lighting power densities. "Lighting power density" describes the density of lighting power in a building or space based on the amount of connected lighting wattage and the space's area, expressed as watts per square foot. ASHRAE 90.1 imposes caps on the maximum allowable lighting power density in building or space types depending on whether the Building Area Method or Space by Space Method is used.

What's new: The lighting power densities in 90.1-2004 have been completely revised. For example, the Building Area Method's LPD values are generally 13-50 percent lower than 90.1-1999, as shown in **Table 4-11**.

Table 4-11. Comparison of lighting power densities between 90.1-1999/2001 and 90.1-2004/2007. Numbers are rounded to nearest whole integer in the % Difference column.

Building Area Method	Lighting Power Densities (W/sq.ft.)		
	1999/2001	2004/2007	% Difference
Automotive Facility	1.5	0.9	-40%
Convention Center	1.4	1.2	-14%
Court House	1.4	1.2	-14%
Dining: Bar Lounge/Leisure	1.5	1.3	-13%
Dining: Cafeteria/Fast Food	1.8	1.4	-22%
Dining: Family	1.9	1.6	-16%
Dormitory	1.5	1.0	-33%
Exercise Center	1.4	1.0	-29%
Gymnasium	1.7	1.1	-35%
Healthcare Clinic	1.6	1.0	-38%
Hospital	1.6	1.2	-25%
Hotel	1.7	1.0	-41%
Library	1.5	1.3	-13%
Manufacturing Facility	2.2	1.3	-41%
Motel	2.0	1.0	-50%

(Continued)

Table 4-11 (*Cont'd*). Comparison of lighting power densities between 90.1-1999/2001 and 90.1-2004/2007. Numbers are rounded to nearest whole integer in the % Difference column.

Building Area Method	Lighting Power Densities (W/sq.ft.)		
	1999/2001	2004/2007	% Difference
Motion Picture Theater	1.6	1.2	-25%
Multi-Family	1.0	0.7	-30%
Museum	1.6	1.1	-31%
Office	1.3	1.0	-23%
Parking Garage	0.3	0.3	--
Penitentiary	1.2	1.0	-17%
Performing Arts Theatre	1.5	1.6	+7%
Police/Fire Station	1.3	1.0	-23%
Post Office	1.6	1.1	-31%
Religious Building	2.2	1.3	-41%
Retail	1.9	1.5	-21%
School/University	1.5	1.2	-20%
Sports Arena	1.5	1.1	-27%
Town Hall	1.4	1.1	-21%
Transportation	1.2	1.0	-17%
Warehouse	1.2	0.8	-33%
Workshop	1.7	1.4	-18%

Exterior lighting power allowances. Outdoor lighting power allowances are dramatically expanded in ASHRAE 90.1-2004 to bring outdoor lighting into the code in the same way that indoor lighting is. The requirements cover 17 typical outdoor applications.

What's New for Lighting in ASHRAE 90.1-2007?

The LPD limits in Standard 90.1-2004 are more stringent than the 1999 and 2001 versions, but 90.1-2007 makes no changes to power allowances. What's new is mostly clarifications and refinements, with a major revision for additional power allowances.

In the Spotlight: Code Confusion

Some important notes about energy codes:

When a new version of a code standard is published, it does not automatically become required for your new construction and renovation projects. Standards and model codes such as ASHRAE 90.1 and IECC are developed by organizations in code-ready language that can be adopted by states and other jurisdictions. When a new standard or model code is published, it is made available to states and jurisdictions to adopt as their energy code if they wish to do so.

Automatic lighting shutoff requirements cannot be satisfied using manual switches. This misperception is based on a typo in 90.1-1999 that has since been corrected.

Manual bi-level switching is not a universal code requirement. It is only required in IECC and some state-specific codes such as California's Title 24. Standard 90.1 does not require it because it is a manual control; instead, 90.1 requires occupancy sensors in many of these spaces.

Alterations do not always require code compliance for every system. While renovations and alterations to existing buildings are covered by energy standards and model codes, the requirements only apply to building systems that are being altered. Lighting that is not affected during a renovation is not subject to compliance. Standard 90.1, for example, does not require that lighting meet code unless 50 percent or more of the luminaires are replaced. (If less than 50 percent are replaced, the new lighting does not have to meet code, but it cannot increase the overall energy consumption of the space either.) Controls do not have to meet code unless the above condition is satisfied plus existing controls are replaced.

Energy codes are intended to apply to all indoor lighting except where specifically exempted. They are not restricted to overhead general lighting.

Below are highlights of several key lighting changes. All comparisons to the previous version of 90.1 refer to 90.1-2004.

Section 9.1.4. This section defines what wattage should be used for luminaires in the installed interior lighting power calculations. Previously for track lighting, if low-voltage track was specified, the maximum wattage of the track system was the transformer wattage, while line-voltage track was penalized with a minimum of 30W/linear foot.

This section now recognizes the wattage of a circuit breaker connected to the system and the wattage of other current-limiting devices.

Track lighting sub-panels enable multiple lengths of track to be applied to one current-limited circuit as well as control of other loads such as low-voltage track and non-track lighting. Track current-limiters are installed at the supply end of each track section; if the connected load exceeds the current limitation, a reset breaker trips.

Section 9.2.2.3. The wattage of furniture-mounted task lighting is now excepted from being included in the installed interior lighting power total as long as it is controlled by automatic shutoff plus either an integral switch or a suitable wall-mounted control device. This means connecting the task lighting to a scheduling panel or occupancy sensor.

Section 9.4.1.3. Exterior luminaires not intended for dusk-to-dawn operation previously had to be controlled by an astronomical time switch. Under 90.1-2007, these luminaires can be controlled by either a time switch or a combination of a time switch and a photosensor. The combined control option provides flexibility and practicality, allowing the user to decide which is best for them.

Section 9.6.2. The biggest changes are in this section, which covers additional lighting power allowance that can be claimed for certain applications, such as retail and decorative lighting, when using the Space by Space Method.

First, language is included that strengthens 90.1's intent that additional allowances can only be used for non-general lighting. Second, the additional lighting power allowed for applications with video display terminals (VDTs) such as computer screens has been discontinued, limiting open and enclosed office spaces to a total 1.1W/sq.ft. And finally, additional lighting power allowances for retail spaces have been revised dramatically.

Standard 90.1-2007 says that separately controlled, non-general lighting installed in sales areas and used to highlight merchandise can claim additional lighting power as shown in **Table 4-12**.

Table 4-12. Additional interior lighting power allowance for sales areas.

If the sales area is used to sell ...	... it is classified as ...	... and the additional interior lighting power allowance is 1000W + ...
All products not listed below	Retail Sales Area 1	Floor Area x 1.0W/sq.ft.
Vehicles, sporting goods, small electronics	Retail Sales Area 2	Floor Area x 1.7W/sq.ft.
Furniture, clothing, cosmetics, artwork	Retail Sales Area 3	Floor Area x 2.6W/sq.ft.
Jewelry, crystal, china	Retail Sales Area 4	Floor Area x 4.2W/sq.ft.

In the Spotlight:
What's Next for Commercial Building Energy Codes

Below are several directions that energy codes appear to be heading in:

- **More daylight harvesting:** The most advanced codes and standards require daylight harvesting. Expect more of this—either mandatory or implemented via credits.
- **Energy focus instead of connected load density:** Energy codes cap allowable lighting power density, but LPD does not recognize how the lighting system is used nor the impact of additional energy-saving technologies such as advanced lighting controls. Expect LPD credits for advanced controls and otherwise a greater emphasis on kWh/sq.ft. instead of W/sq.ft.
- **Demand response:** Currently, only the most advanced codes and standards include demand response. As managing peak demand is a critical concern in some regions, expect demand response to grow in importance as a mandatory requirement.
- **Existing building coverage:** Future codes are likely to address existing buildings. ASHRAE 90.1-2010, still in development at the time of writing, is expected to cover both indoor and outdoor lamp/ballast retrofits and all luminaire replacement, with the lighting in the space required to comply with the LPD and automatic shutoff requirements of the Standard. The exception would be alterations that involve less than 10 percent of the connected lighting load, as long as those alterations do not increased the installed LPD.
- **Green buildings:** Currently, LEED is being used as a de facto code, with various jurisdictions requiring certain buildings to achieve a degree of LEED registration. With the development of ASHRAE 189.1 (which will be followed by California's CAL Green in 2011 and ICC's IgCC in 2012), green building code standards are becoming available and may be adopted by states. These codes go far beyond energy codes by requiring higher levels of energy efficiency and also achievement of other sustainability targets, such as water efficiency, etc.
- **Residential coverage:** Most residential energy codes focus on HVAC. IECC 2009, however, requires a portion of the permanently installed lamps in new homes to be high-efficacy. Expect more coverage of lighting in residential energy codes.
- **Stricter and stricter:** At the time of writing, legislation was pending in Congress that was highly publicized for cap and trade rules, but also required DOE to establish a national energy standard based on aggressive savings over current code standards. As a result, expect energy codes to become more and more restrictive.

LIGHTING AND LEED 2009

Sustainable construction represents about 10 percent of the current commercial and institutional building market, according to McGraw-Hill, which predicts that demand will increase to 20- 25 percent by 2013, or $96-140 billion.

According to the U.S. Green Building Council (USGBC), creators of the Leadership in Energy and Environmental Design (LEED) "green" building rating system, sustainable construction reduced the nation's electric bill by $1.3 billion between 2000 and 2008, and will save another $6 billion from 2009 to 2013.

LEED has provided focus to the sustainable design movement and, to a significant extent, put a face on it.

In LEED, the sustainable design movement has identified strategies to address multiple criteria that define a green building. Regarding lighting, LEED has increased the use of the most energy-efficient lighting technologies, collaboration between architects and lighting designers particularly on daylighting, and the use of supplemental task lighting and personal lighting control.

LEED 3.0, the latest version of LEED, was launched April 27, 2009. By June 27, 2009, all LEED projects had to begin registering under the new LEED 2009 rating system.

This version of LEED incorporates all of the rating systems that address commercial buildings—New Construction, Commercial Interiors, Schools and Existing Buildings into three LEED 2009 systems: Green Building Design & Construction, Green Interior Design & Construction and Green Building Operations & Maintenance. While the majority of individual credits is familiar, LEED 2009 includes significant changes.

First, LEED credits are now weighted, resulting in more emphasis on energy and atmosphere points. A number of credits are updated. And bonus points are awarded for LEED credits considered to be a priority for particular regions.

Beyond daylighting, the biggest impact lighting can have on LEED points is optimized energy performance, light controllability, light pollution and mercury content in lamps.

Energy Performance

LEED 2009 requires that the building demonstrate a 10 percent minimum energy reduction compared to an ASHRAE 90.1-2007 (or California

Title 24) compliant building. Between one and 19 credits are available based on going further than that, with one point being equal to about a two percent reduction. This may involve highly efficient lamps, ballasts and luminaires; innovative design; and advanced controls such as daylight harvesting dimming. Additionally, commissioning is required for building controls such as lighting controls, and enhanced commissioning and measurement and verification are rewarded with up to five LEED points.

Light Controllability

To promote occupant productivity and well-being, LEED 2009 offers credit for controllable lighting. A point is available for providing individual lighting controls for at least 90 percent of the occupants (50 percent for LEED 2009 Green Building Operations and Maintenance) as well as lighting controllability for all shared multi-occupant spaces (i.e., classrooms,

Figure 4-3. The Energy Foundation utilizes energy-efficient lighting and an advanced digital lighting control system to dramatically reduce energy costs while contributing to a LEED for Commercial Interiors Platinum rating. Photo courtesy of Lutron Electronics.

conference rooms). Dimmers and multiscene lighting controls that allow for adjustable lighting levels to meet occupant needs and preferences help achieve the requirements for this credit.

In addition, LEED 2009, Green Interior Design & Construction offers up to 3 points for the appropriate usage of daylight harvesting controls and occupancy sensors. Two points are available for using daylight controls: 1) in all daylighted areas (1 point) and 2) on 50 percent of the connected lighting load (1 point). And one point available for using occupancy sensors on 75 percent of connected lighting load.

Light Pollution

The light pollution credit requires that the building minimize light trespass and skyglow, and is based on the lighting zone where the project is located. Exterior areas that are rural or concerned with stringent control of nighttime lighting (Zones 1 and 2) must be designed to a lower light level with greater control of the light emitted directly into the sky or outside of property lines. Lighting in all zones must not exceed ASHRAE 90.1-2007 (with errata) power density limits.

Typical solutions will use lower-wattage sources and luminaire optics that have no uplight and good backlighting control. For area lighting in a parking lot, a 150-250W fully shielded shoebox-type luminaire can be a good solution. For lighting near the property perimeter, a sharp-cutoff, forward-throw optic will minimize light trespass. When decorative or landscape lighting is needed, luminaires with low-wattage sources that minimize uplight should be used. Landscape lighting should also use low-wattage sources while aiming light in a downward direction whenever possible.

Additionally, for interior areas, LEED 2009 requires that light emission through building windows be minimized during the night hours (11 PM to 5 AM). Light must not be allowed to exit windows or lighting power must be reduced by at least 50 percent. Automatic window shades and automatic shutoff controls can help meet this requirement.

Mercury Content in Lamps

LEED 2009, Green Building Operations & Maintenance reduces the number of points available for mercury reduction from two to one, which is achieved for achieving 90 picograms per lumen-hour for lighting used in the building (based on a weighted average of 90 percent of the lamps). The lamp manufacturers offer calculators that can help with this. Lamp recycling is also required: The owner must have a lamp recycling contract in place, demonstrated by proper documentation, to receive accreditation.

Caveat Emptor

LEED 2009 places greater emphasis on energy efficiency than previous versions and as a primary energy user lighting can play a significant role in reducing building energy consumption. However, while an extremely low lighting power density may be valued as "good" in that it is rewarded with LEED points, designers should be cautious about ensuring that the lighting system, however efficient, achieve the space's lighting quality goals first. LEED should be used as a tool to achieve a green building, not the sole consideration.

Figure 4-4. The USGBC's 22,000-sq.ft. office space utilizes a lighting design by Bliss Fasman Lighting Design that combines high efficiency, daylight and views, and LED task lighting to contribute to a LEED for Commercial Interiors Platinum rating. Photo by Prakash Patel.

GREEN BUILDING CODES

ASHRAE 189.1, *Standard for the Design of High-Performance, Green Buildings Except Low-Rise Residential Buildings,* published in January 2010 by ASHRAE, provides the first code-intended commercial green building standard in the United States.

Released jointly with IES and USGBC, ASHRAE 189.1 provides criteria by which a building can be judged as being "green"—that is, built and operated to a high standard of design sustainability. As with the ASHRAE 90.1 energy standard, 189.1 is not a code, although it can be adopted by governments as the basis for codes. It is also not a rating system such as LEED.

ASHRAE 189.1 is likely to be adopted as a code for certain types of construction in jurisdictions currently requiring LEED certification. While the most ambitious policies require LEED certification for some private commercial buildings, LEED is generally required for public construction in these jurisdictions.

ASHRAE 189.1 covers the same building types as Standard 90.1 and includes the same building areas as LEED, including energy efficiency, site sustainability, water use efficiency, indoor environmental quality and the building's impact on materials, resources and the atmosphere. The energy section is based on the 90.1 energy standard, but is more efficient than 90.1-2007, requires energy measurement for verification purposes, contains renewable energy provisions, and requires the capability to reduce peak electric demand. Overall, the goal is to achieve at least a 30 percent reduction in energy cost over 90.1-2007.

As a major building system, lighting plays a major role in ASHRAE 189.1, covered in Section 7.4.6. This section, in turn, is based on Section 9 of ASHRAE 90.1 with several significant additions and modifications to increase energy savings. The lighting power densities in 90.1, for example, are capped at 90 percent in 189.1. Most of the rest of the changes involve more aggressive, mandatory use of automatic lighting controls.

The installation of occupancy sensors, for example, is mandated in a broader range of spaces, and manual-ON operation is required unless an alternative is specifically allowed. ASHRAE 189.1 identifies a number of enclosed spaces where occupancy sensing is required, such as classrooms, small offices and conference and meeting rooms. In covered spaces, the sensor must provide manual-ON/automatic-OFF operation, or instead of manual ON, the sensor can combine with a multilevel switching scheme and turn the lights ON automatically to a lower light level, with higher light levels requiring manual switching. ASHRAE 189.1 also requires occupancy sensors in spaces that often require continuous lighting, such as commercial and industrial storage stack areas, library stack areas and hallways in hotels, motels, dorms and multifamily buildings, with an exception for HID lighting with a low power density. In these spaces, the sensor must be installed with a multilevel switching or dimming system that reduces lighting power by at least 50 percent when they are unoccupied.

Egress and security lighting that must be operated 24/7 is capped at 0.1W/sq.ft., but additional lighting can be installed as long as it is controlled by an automatic shutoff device.

Continuous dimming or stepped switching automatic daylight harvesting controls are required for daylight zones, under skylights or next to vertical fenestrations, that combine in each enclosed space to exceed 250 sq.ft. These daylight zones are defined at the beginning of ASHRAE 189.1. Daylight harvesting is beginning to be required by other standards such as the 2009 IECC, Title 24 and the upcoming 90.1-2010, but ASHRAE 189.1 is aggressive in that it requires automatic lighting control instead of merely separate zoning. There are many exceptions to this requirement, such as window display and exhibition lighting, hotel guest room lights and conference room lighting operated by local dimming controls.

Outdoor lighting must comply with Section 9 of ASHRAE 90.1, but some applications—parking lots, building facades, garages and others— require either a motion sensor and photocontrol or automatic controls that reduce lighting power by at least 50 percent, with few exceptions.

Additionally, two other provisions of interest include submetering and demand response. Section 7.3.3.1 requires submetering of lighting if the connected load is greater than 50 kVA. Section 7.4.5.1 requires the building to contain automatic systems capable of reducing peak electric demand by at least 10 percent, not including standby power generation, which may involve automatic shutoff and dimming of lighting systems.

California recently passed CAL Green, a state green building code, which will apply to new commercial buildings starting January 2011. And the International Code Council, creator of the IECC, is expected to announce publication of the International Green Construction Code (IgCC) standard in 2012, when the next version of IECC will be released.

Chapter 5

Evaluating Lighting Equipment

There are no bad products, only bad applications.

—Howard Brandston

EVALUATING LIGHT SOURCES

After identifying the lighting needs and making decisions about what surfaces we want to light and with what intensities, we can select our lighting system. A typical lighting system includes the light source (lamp), a ballast if the lamp needs one; a luminaire to house the lamp/ballast, connect it to the power supply, and control the direction and intensity of the light; and a control device for dimming the light or turning it on or off.

At the heart of this system is the light source, our first decision point.

All lamps include one or two bases for connection to the power supply, a glass bulb, gases inside the bulb, and a light-producing element such as a filament. How the lamp actually produces the light has given rise to a number of different types of lighting: fluorescent, incandescent, high-intensity discharge (HID) and solid-state lighting (SSL), including light-emitting diode (LED) lighting. Each of these major families has numerous variations on the theme. Each family, and each variation, makes the lamp suitable for different lighting needs. Incandescent and halogen lamps, for example, are more common in museums and retail stores, fluorescent lamps are common in offices and schools, HID lamps are common in parking lot and streetlighting, and LED lighting has become competitive in directional applications such as downlights and undercabinet lights.

Over the past two years, this book's author has been honored to work with Howard Wolfman, PE, principal of Lumispec Consulting, to present the LIGHTFAIR What's New in Lamps and Ballasts seminar. The industry maintains a brisk pace of innovation driven by energy efficiency, new applications and designer preferences. Recent examples include energy-saving 49-51W replacements of 54W T5HO lamps with no loss of light output, extended-life T8 lamps providing a rated life of up to 40,000+ hours at 12 hours/start on an instant-start ballast, T5VHO and T8VHO lamps producing 7,200 lumens of initial light output, new NEMA Premium fluorescent ballasts, new load-shedding and light level switching fluorescent ballasts, new high ballast factor fluorescent ballasts, new electronic HID ballasts, ceramic metal halide lamps available as small as 15W, and energy-saving 23-30W fluorescent T8 lamps that are also dimmable. To name a few!

Selection characteristics include light output, distribution, input watts, efficacy, intensity, rated service life, size, surface brightness, color characteristics, electrical operating characteristics, requirement of additional equipment such as ballasts, compatibility with the electrical system, suitability for the operating environment, compatibility with frequent switching devices such as occupancy sensors, ability to be dimmed, energy efficiency and economics.

Below are several key questions one should ask about a light source. The answers are tools for designers wishing to control how objects and surfaces appear in the space.

What is the distribution of the light?	Candela	cd
How long does the lamp last?	Life (hours)	hrs
How much light does it produce?	Light output (lumens)	lm or L
How much electricity does the system require?	Wattage (watts)	W
How efficient is it compared to others?	Efficacy	lm/W
What is the color appearance of the source?	Color appearance	K
How well does the light source render colors?	Color rendering	CRI

What is the Distribution of the Light?

Most light sources have a particular standardized size and shape that affects how objects and surfaces will be illuminated.

Dear Reader:
Note that lighting technology continues to progress at a fast pace, particularly concerning development of solid-state lighting such as LED. As a result, by the time you read this at some point after the time of writing, some of the data in this chapter may be outdated. Please consult the manufacturers of lamps, ballasts, controls and luminaires on the latest developments in their products.

Table 5-1. Performance characteristics of major lamp types. Source: LIGHTFAIR 2009 Lamp and Ballast Basics Seminar by Craig A. Bernecker, PhD, FIES, LC.

Type	Warm-up Time (Minutes)	Re-strike Time (Minutes)	Temperature Effects	Dimmable?	Ballasted?
Incandescent	NA	NA	NA	Y, down to 0%	N
Halogen	NA	NA	NA	Y, down to 0%	N
Fluorescent	NA	NA	Starting Output	Y, down to 1% using dimming ballast	Y
Mercury Vapor	5-7	3-6	Starting	Y, down to 50%	Y
Metal Halide	2-4	10-15	Starting	Y, down to 50%	Y
High Pressure Sodium	3-4	1	Starting	Y, down to 50%	Y

Table 5-2. Performance characteristics of major lamp types. Source: LIGHTFAIR 2009 Lamp and Ballast Basics Seminar by Craig A. Bernecker, PhD, FIES, LC.

Type	Watts (W)	Efficacy (lumens/W)	Lamp Lumen Depreciation	Life (hours)	Correlated Color Temperature (K)	Color Rendering Index (CRI)
Incandescent	0.1 - 1500	15-25	0.90	600-4000	2700	90-95
Halogen	0.5-1500	20-35	0.95	2000-6000	2900	90-100
Fluorescent	5-215	74-100	0.92-0.66	12000-20000	3000-6000	50-90
Mercury Vapor	35-1000	20-63	0.84-0.55	20000+	3000-6000	20-50
Metal Halide	35-1500	80-125	0.92-0.59	7500-20000	3000-4500	60-70
High Pressure Sodium	35-1000	65-140	0.92-0.90	20000+	2000-3000	20-30

Point sources are small lamps, often featuring a clear outer glass bulb revealing the arc tube or bare incandescent filament, used to produce dramatic highlights and pronounced shadows (see **Figure 5-2**).

Linear sources, such as tubular fluorescent lamps, emit diffuse output from the surface of the lamp, softening shadows (see **Figure 5-3**).

Area sources are not lamps but instead large surfaces that emit diffuse light, such as ceilings reflecting illumination from an indirect luminaire, or the surface of a luminous bowl pendant luminaire (see **Figure 5-4**).

In the Spotlight: Higher Efficiency, Longer Life, Controllability

In recent years, lamp manufacturers have made significant gains in efficiency and service life in linear lighting by adjusting the composition of gases and fill pressure inside the lamp. By tuning composition and fill pressure, power, life and light output can be optimized.

For example, a T8 lamp can be designed with the same life rating but with a higher level of efficiency, giving us the energy-saving T8 family consisting of 25W, 28W and 30W energy-saving T8 lamps. This is achieved by adding a higher percentage of krypton to the lamp's internal gas mixture, resulting in a lower operating voltage which, in turn, reduces lamp power. In 2010, Sylvania announced a new member of the family: 4-ft. 23W T8 lamps, which produce about 30 percent less light output than a standard T8 while producing 28 percent energy savings, marketed for use in coves, corridors and overlighted existing office applications. Sylvania also announced that its 23W and also 25W, 28W and 30W T8 lamps are dimmable.

Consumption of T5 lamps continues to grow in applications such as indirect lighting, some direct luminaires and high-bay replacement luminaires. In this category, the major lamp manufacturers are launching energy-efficient T5HO lamps that deliver the same light output as 54W T5HO lamps but for a lower wattage, increasing efficacy from 93 to as high as 102 lumens/W. Examples include Philips 49W T5HO lamps and GE Watt Miser 51W T5HO lamps. Sylvania also offers Pentron SuperSaver 47W T5HO lamps, which provide about 13 percent energy savings for about 9 percent less light output.

Alternately, in both T8 and T5 lamps, increasing the fill pressure can extend lamp life by preserving the cathodes. For example, the traditional 32W T8 can be designed at longer life—36,000 to 42,000 hours—with the same light output. Life can be extended to 46,000 hours with minimal light loss at 10 hours per start on a programmed-start ballast. LEDs continue to advance steadily, but fluorescent is clearly not going to go down without a fight.

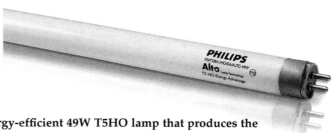

Figure 5-1. Energy-efficient 49W T5HO lamp that produces the same light output as a 54W T5HO lamp but for 9 percent less lamp power. Photo courtesy of Philips Lighting Company.

Figure 5-2. Geometric objects lighted by a point source. Photo courtesy of Peter Ngai.

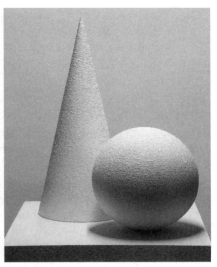

Figure 5-3. Geometric objects lighted by a linear source. Photo courtesy of Peter Ngai.

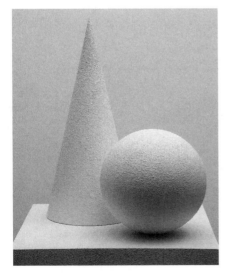

Figure 5-4. Geometric objects lighted by an area source. Photo courtesy of Peter Ngai.

How Long Does The Lamp Last?

Lamp life is expressed in operating hours, or "burn time," as rated by its manufacturer. You can find a given lamp's rated life in its manufacturer's catalog.

Rated life and actual life may not be exactly the same. The manufacturer rating is based on a standardized test method, while actual life is influenced by field conditions. High line voltage, connecting the lamp to the wrong type of ballast, and other problems can shorten lamp life.

Fluorescent and HID rated lamp life is an average that is predictable for a large population of lamps. At 100 percent of rated life,

50 percent of a large group of lamps can be expected to have failed. The rate of failure is shown on the lamp's *mortality curve* (see **Figure 5-5**).

For example, if we have 100 lamps and their rated life is 20,000 hours, then at 20,000 hours, 50 of the lamps can be expected to have failed. If we look at the mortality curve for this lamp, we see that at 70 percent of their rated life, only 7 percent have failed. After 70-80 percent of lamp life, the rate of failure accelerates dramatically.

For gaseous discharge lighting systems (fluorescent, HID), rated lamp life is based on *hours per start*. Fluorescent lamps are typically rated at 3 hours per start and HID lamps are typically rated at 10 hours per start, for example, with a cycle being ON for a number of hours, then OFF for 15-20 minutes. A fluorescent lamp rated at 20,000 hours at 3 hours per start may be rated at 26,000 hours at 10 hours per start. If the lamp is turned ON and never turned OFF, it could last as long as 30,000 hours or longer. The latest generation of "extended life" fluorescent lamps can achieve a rated life 46,000 hours at 12 hours per start.

For fluorescent lamps, lamp life also depends on the starting method. For example, an average T8 lamp is rated at 20,000 hours at 3 hours per start on an instant-start ballast (the most popular electronic ballast) and 24,000 hours at 3 hours per start on a programmed-start ballast.

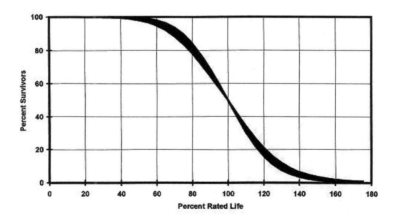

Figure 5-5. Typical mortality curve for fluorescent lamps. Large groups of lamps tend to follow this curve. Graphic courtesy of the interNational Association of Lighting Management Companies (NALMCO).

Table 5-3. Extended-life T8 lamp life.

	Instant Start Ballast		Programmed Start Ballast	
	3 hrs/start	12 hrs/start	3 hrs/start	12 hrs/start
Industry Average (32W T8)	20,000	24,000	24,000	30,000
"XL" Lamps (25-32W T8)	21,000-36,000	30,000-40,000	30,000-40,000	36,000-46,000
% Greater Rated Life	5-80%	25-67%	25-67%	20-53%

Lamp burnouts (a "burnout" simply defined as a lamp that has expired) are a light loss factor that should be accounted for in determining the number of lumens, lamps and luminaires required to produce the desired maintained light level. Generally, there are two ways to treat burnouts: replace them immediately or after a specified period of time. The lamp burnout factor is expressed as (1 - % Lamps Allowed to Fail Without Being Replaced). If burnouts are spot relamped promptly upon failure or group relamped prior to any failure, the burnout factor would be 1. If some 5 percent of the lamps are burnouts at any given time, then the lamp burnout factor would be 0.95. Lighting designers often assume that burnouts will be replaced promptly unless the luminaires are difficult (costly) to reach and/or some lamps are strategically allowed to fail before group relamping, which is typically recommended at about 60-70 percent of rated life.

Aside from burnouts, another failure mode is lumen depreciation, used to determine LED and induction lamp life. For these sources, rated life is determined as the point in time when light output is expected to decline below an acceptable limit—typically 70 percent of initial rated light output, called the light source's L_{70} life rating.

The owner may consider the lamp as having "failed" as a result of other factors, such as unacceptable color shift at end of life, lamp efficacy falling to a point where it is no longer economical to continue operating the lamp, lamp cycling (high-pressure sodium lamps) or instabilities (all gaseous discharge lamps), and the point at which it becomes economical in terms of labor to group relamp rather than continue spot relamping individual failures.

In the Spotlight: Linear Amalgam Lamps

Amalgam technology, fairly common among plug-in compact fluorescent lamps, is now available in linear T5HO, T5VHO and T8VHO fluorescent lamps, making fluorescent lighting competitive in many traditional HID applications (see **Figure 5-6**).

Fluorescent lamps produce visible light through a gaseous discharge process called fluorescence: A current is passed through a lamp. The current electrically excites mercury atoms to release ultraviolet energy. The ultraviolet energy is converted into visible light by a phosphor powder coating the lamp's inner bulb wall.

Mercury pressure inside the lamp is regulated by a "cold spot" created by placing the filament at one end of the lamp deeper than the other. If the mercury pressure is too low, not enough UV energy is generated by the discharge. If the mercury pressure is too high, UV energy is reabsorbed by mercury atoms in the lamp. Both mean less light output is being emitted.

Since typical operating temperature of the cold spot is slightly higher than the ambient temperature surrounding the lamp, a lamp's actual light output in the field is dependent on ambient temperatures around the lamp.

This relationship is so clear that rated light output in lamp catalogs is based on a specific design temperature: 25°C for compact fluorescent and linear T8 and T12 lamps, and 35°C for linear T5 and T5HO lamps. Since maintaining an exact ambient temperature is not practical, we are more concerned about getting most of the light output over a range. For example, CFLs produce greater than 90 percent of their initial rated light output over a 10-45°C ambient temperature range, while T8VHO lamps deliver it over a 15-40°C range and T5HO and T5VHO lamps deliver it over a 25-50°C, depending on the manufacturer and product. Beyond these ranges, however, light output begins to suffer significantly. Since light output falls but input wattage does not, efficiency, expressed in lumens/W, also drops dramatically.

In many typical commercial building spaces such as offices, temperature is tightly regulated and so this is generally not a significant issue. But in spaces where HID lamps are used, such as indoor spaces where temperature is not conditioned as well as outdoor spaces, ambient temperatures can get much hotter and colder than the design rating.

That's where amalgam technology proves its value.

Amalgam technology, a part of CFL designs for the past 15+ years, is now becoming increasingly available in linear fluorescent lamps intended for applications dominated by HID lighting. In an amalgam lamp, a different approach is used to control mercury vapor pressure. A metal is added to the lamp, which alloys with the liquid mercury and creates an amalgam. The result is the lamp is able to produce greater than 90 percent of its initial rated output over a much broader range than non-amalgamated lamps.

Figure 5-6. Amalgam technology enables linear fluorescent lighting to extend into applications that get as cold as 5-10°C and as hot as 65-75°C, such as unconditioned high-bay spaces, garages, even outdoor applications. Courtesy of Philips Lighting Company.

Amalgam CFLs, for example, available in 9-26W self-ballasted CFLs and 13-70W plug-in lamps, with 26W, 32W and 42W plug-in CFLs being most common, can deliver greater than 90 percent of initial rated light output over a 5-65°C ambient temperature range. Amalgam T5HO, available in 54W 4-ft. lamps, and T5VHO, available in 95W and 120W 4- and 5-ft lamps, can deliver greater than 90 percent of initial rated light output over a 5-7°C to 70-75°C range, depending on the manufacturer. And T8VHO lamps, available in 84W 4-ft. lamps, can deliver it over a 10-65°C range.

As one would expect, ideal applications include spaces where fluorescent lighting is more efficient than HID, but where strong temperature fluctuations occur during operating hours. These applications include warehouses and distribution centers, hangars and industrial facilities. Another strong application is sealed luminaires in walk-in freezer applications. And a potential application that has yet to become fully realized is outdoor lighting, with potential to compete with HID lighting in outdoor area lighting, signage, gas station, canopy and accent lighting.

Note that linear amalgam fluorescent lamps may be slower to start and respond to dimming signals than non-amalgam lamps, which may limit energy savings potential using automatic controls. According to manufacturers, amalgam lamps can take up to three minutes to achieve full brightness, which may limit application with occupancy sensors (lamps extinguished for short periods of time, however, usually return to full brightness very quickly). Amalgam lamps are also slower to react to dimming signals and otherwise impose limitations on dimming range, which may limit application with automatic dimming controls. If a slight delay in dimming can be tolerated, dimming to 50 percent of full light output may be practical as a means of saving energy in direct response to daylight or lack of occupancy, however.

With the growing availability of amalgam technology for linear lamps, building owners now have a choice of lighting for applications that were once considered exclusively the domain of the HID lamp.

How Much Light Does it Produce?

Lamps emit light measured in lumens; the initial light output of a given lamp can be found in its manufacturer's catalog. A typical T8 fluorescent lamp, for example, produces about 2900 lumens. Fluorescent and HID lamps are operated on ballasts, which may affect the light output through the application of *ballast factor*.

As the lamp ages and nears end of life, however, it produces less and less light than it did when it was new. Several factors cause this phenomenon, including deposits inside the glass bulb wall and deterioration of the phosphor coating on the inside of the bulb. This *lamp lumen depreciation* is characteristic of all lamps, although they vary in extent of depreciation (see **Figure 5-7**). It is visualized on a *lumen maintenance curve*, which plots light output versus time. This data can be very useful, while keeping in mind that manufacturers present optimal data.

Lamp lumen depreciation (LLD) is also a light loss factor included in calculations to determine the design light level. If the lighting system is spot relamped, the LLD factor is the decimal fraction of initial light output that is produced at mean rated life. Expressed as a percentage, it gives us the lamp's *lumen maintenance* value. If the lighting system is group relamped, the LLD factor is calculated at the group relamping interval. Since we are primarily concerned with maintained light levels, initial

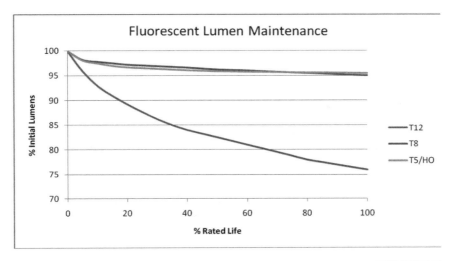

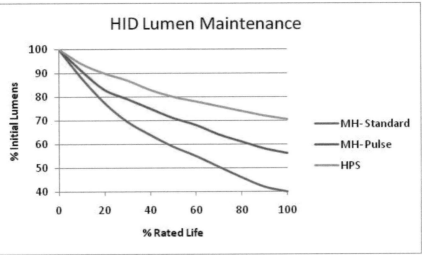

Figures 5-7a and 5-7b. Lamp lumen depreciation curves for popular fluorescent (a) and HID (b) light sources. MH = Metal Halide. HPS = High-Pressure Sodium. Graphic courtesy of the New Buildings Institute.

rated light output for a given light source has practical value only as a starting point in design calculations, with the lamp's lumen maintenance being an important additional factor.

As an example of the importance of lumen maintenance, consider a high-bay installation consisting of 400W probe-start metal halide luminaires, which we are considering replacing with fluorescent 6-lamp high-performance T8 (Super T8) fluorescent luminaires for 50 percent energy savings. At first glance, this does not seem to be a wise replacement. With a ballast factor of 1.0, each metal halide luminaire produces 36,000 initial lumens. The fluorescent luminaire, with a selected ballast factor of 1.18, produces nearly 22,000 lumens. The energy savings are good, but how can this replacement produce comparable light levels?

The answer, of course, is lumen maintenance. Probe-start metal halide lamps experience a higher level of lumen depreciation than T8 lamps. For example, a 400W metal halide lamp is expected to lose up to 35 percent of its light output at 40 percent of life, while a T8 lamp is expected to lose only about 5 percent (see **Table 5-4**). The 400W lamp's efficacy drops from about 80 lumens/W to about 50 lumens/W, marking it as a relatively inefficient light source compared to the Super T8, which has a maintained efficacy of 94 lumens/W. As a result, a 6-lamp Super T8 lamp-ballast system produces about 10 percent fewer mean lumens for about 50 percent less electrical energy.

Table 5-4. Comparison of fluorescent and 400W probe-start metal halide high-bay luminaires. Data source: Philips Lighting Electronics.

System	Initial Lumens	Lumen Maintenance	Mean Lumens @ 40% Lamp Life**	Relative Mean Lumen Output
400W Probe-Start Metal Halide	36,000	65%	23,400	100%
4-Lamp T5HO Fluorescent	20,000	95%	19,000	81%
6-Lamp T5HO Fluorescent	30,000	95%	28,500	121%
6-Lamp Super T8 Fluorescent	21,948	95%	20,851	89%

**Fluorescent lamp lumens are based on optimal temperatures; adjust as needed.
**To further the comparison, consider researching and comparing these numbers at end of lamp life rather than at the mean.

How Much Electrical Energy Does the System Require?

In review, a watt is a unit of measurement of required electric power. Rated input wattage for any electrical equipment is the amount of power it requires to operate at any given instant of time. The input watts for a given lamp can be found in its manufacturer's catalog. If the lamp is a gaseous discharge lamp (fluorescent or HID), however, this rating is only nominal, however. The lamp requires a ballast. These two components operate as a system for which we must account for electrical losses. As a result, the rated system wattage for the given lamp-ballast combination, which can be found in the ballast manufacturer's catalog, offers greater practical value.

In The Spotlight: NEMA Premium Ballasts

In recent years, ballast manufacturers have begun marketing a premium fluorescent ballast called a "high-efficiency ballast" for operation of 4-ft. T8 lamps.

A high-efficiency ballast provides the same level of light output as a standard electronic ballast, but does it more efficiently; these ballasts operate at over 90 percent efficiency and in some cases up to 95 percent, providing 2-5W per ballast savings compared to standard systems. For a cost adder of about 10-20 percent, the ballast can generate an energy savings adder of up to 5-7 percent (see **Table 5-5**).

Table 5-5. High-efficiency electronic ballasts can produce additional 5-7 percent energy savings in retrofits. Source: Sylvania.

Model	Ballast Factor	Lamp/Ballast Input Power		Savings	
		Standard Electronic Ballast	High-Efficiency Electronic Ballast	Power	Percent
(1) T8 lamp	0.88	30W	28W	2W	7%
(2) T8 lamps	0.88	59W	55W	4W	7%
(3) T8 lamps	0.88	86W	82W	4W	5%
(4) T8 lamps	0.88	112W	107W	5W	5%

For example, a lighting system consisting of a 0.77 ballast factor ballast operating two T8 lamps may draw 51W with a standard ballast and 47W with a high-efficiency ballast, resulting in added savings of 4W.

One manufacturer averages savings to around a dollar per lamp per year. Doing some quick math, that would be a lamp operating 2,500 hours per year (10 hours per day, five days per week, 50 weeks per year) at a cost of $0.10/kWh. Reasonable.

Taking this one step further, suppose we install two high-efficiency ballasts driving four T8 lamps in luminaires mounted on 10-ft. x 10-ft. centers (100 sq. ft. area). The use of high-efficiency ballasts would add about $0.03-$0.06 per sq. ft. to the cost of the project while reducing annual energy costs by about $0.04 per sq. ft., generating a simple payback of 8-16 months. Such combinations can reduce system watts by a total of more than 45 percent when upgrading older T12 systems and 20-30 percent when replacing standard T8 systems.

High-efficiency ballasts are therefore popular for retrofits because they offer a relatively low-risk way to increase energy savings while providing light levels equivalent to a typical 34W T12 system. The ballasts are suitable for almost any commercial application that is suited for 4-ft. T8 lamps operated by standard electronic ballasts.

The ballast can be specified as instant or programmed start. High-efficiency programmed-start ballasts, which can offer the same efficiency as standard instant-start ballasts, reduce wear and tear on the lamp that occurs during startup, which can maximize lamp life in frequently switched applications such as occupancy sensor installations.

The ballast can be specified as low (<0.86), normal (0.86-1) and high (>1) ballast factor, providing flexibility of selection of light levels and utility in a broad range of applications, including hi-bay spaces. At LIGHTFAIR 2010, Sylvania introduced a two-lamp high-efficiency ballast with a medium ballast factor of about 1.0. Many ballasts are available that offer universal-voltage operation, simplifying installation for contractors. Both fixed-output and dimmable models are available. And some ballasts are available with features such as anti-striation (eliminates "barber-poling" effect) and anti-arcing, which enhance safety and reduce concerns about maintenance.

A major problem with high-efficiency ballasts, however, has historically been a clear definition of just what this type of ballast is, and how to recognize one easily among all of the marketing clutter in which every electronic ballast is called a "high-efficiency" product.

To provide clarity and formally recognize the industry's most efficient ballasts for 4-ft. T8 lamps, NEMA launched the Premium Ballast program in 2008. The program, based on the successful NEMA Premium model for electric motors and transformers, creates an efficiency target established by Consortium for Energy Efficiency (CEE) specifications and recognizes ballasts that meet it with a special logo that can be printed on the ballast label (see **Figure 5-8**).

Figure 5-8. The NEMA Premium Ballast mark recognizes the most-efficient electronic ballasts for 4-ft. fluorescent T8 lamps.

As of the time of writing, Acuity Brands Lighting, American Ballast, Espen, GE, Halco, Keystone Technologies, Leviton, McWong International, Philips, Robertson Worldwide, Sunpark Electronics, Sylvania, Technical Consumer Products and Universal Lighting Technologies have all certified high-efficiency products as NEMA Premium Ballasts. NEMA has published a list of these products in a PDF document titled "NEMA Premium Electronic Ballast Program" posted at www.nema.org/gov/energy/efficiency/index.cfm.

As one might expect, this category is subject to significant innovation. A significant breakthrough by the major ballast manufacturers over the past 12-18 months is parallel lamp operation for programmed-start high-efficiency electronic T8 ballasts, in addition to series-wired models. Parallel lamp operation provides independent lamp operation within the luminaire. When one lamp reaches its end of life, the remaining lamps stay lit (see **Figure 5-9**). As

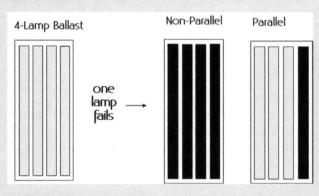

Figure 5-9. A significant breakthrough by the major ballast manufacturers over the past year is parallel lamp operation for programmed-start high-efficiency electronic T8 ballasts in addition to series-wired models. Parallel lamp operation provides independent lamp operation within the luminaire. This feature is expected to increase application of programmed-start ballasts, which are more friendly to use with occupancy sensors.

a result, fewer lamps are replaced because the end-user can tell which lamp has reached end of life. With more lamps remaining ON, it reduces the urgency of relamping and preserves light levels, uniformity and appearance of the lighting system. Examples include Universal's Ultim8, Sylvania's PROStart and Philips' Optanium. This feature is expected to increase application of programmed-start ballasts, which are more friendly to use with occupancy sensors. GE has a similar product in its UltraStart family that also allows 0-10V dimming to 3 percent of light output, with normal and high ballast factor providing a choice of full light output.

GE's dimmable ballast is part of another trend, which is growing friendliness of ballasts with switching and dimming. Universal's Ultim8 ballast, for example, is available in programmed-start models that start the lamps in about half a second, suitable for occupancy sensor installations such as warehouses where fast-moving lift trucks are common. Philips' EssentiaLine 0-10V dimming high-efficiency ballast provides a simple and highly affordable approach to dimming, making it more accessible by reducing the cost of the ballast. And light level switching ballasts, such as Slvania's Quickstep Bilevel T8 Switching ballast and GE's instant-start load-shed dim and bilevel switching ballasts, are becoming more popular for demand response and bilevel switching applications.

NEMA advises the following specification language for retrofit or spot replacement: "Ballast shall be NEMA Premium electronic ballast (do not substitute)." Then specify the starting method, number of lamps (1-4) and ballast factor.

The NEMA Premium Ballast program currently covers electronic ballasts for operation of 4-ft. T8 lamps, but may expand in the future to also cover T4 and T5 ballasts, HID ballasts and LED drivers and power supplies.

How Efficient is it Compared to Others?

The relative efficiency of lamps and lamp-ballast system combinations can be compared using a simple metric called efficacy. Efficacy, analogous to miles-per-gallon ratings for vehicles, expresses the ratio of light output per unit of electrical input, or lumens/W.

Efficacy declines over time, as lamps exhibit lamp lumen depreciation but maintain constant wattage; for this reason, maintained efficacy has more practical value in decision making compared to initial efficacy. Similarly, note that any field condition that affects light output but not wattage, such as ambient temperatures, will affect efficacy. Additionally, note that if the lamp is to be dimmed, efficacy may gradually decline over the lamp's dimming range.

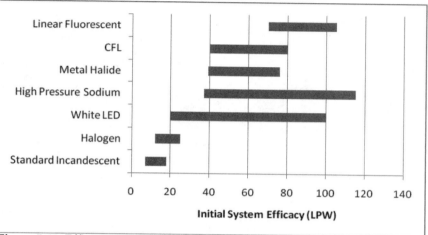

Figure 5-10. Efficacy comparison of light sources for general lighting. Graphic courtesy of New Buildings Institute.

While considered a very useful tool for comparing the relative efficiency of light sources, lighting systems and luminaires, note that efficacy has its limitations. First, it is truly useful only when comparing lamps that have the same light distribution, color characteristics and general configuration. Second, it does not account for the overall effectiveness of the product. Kevin Willmorth, writing in the January/February 2009 issue of *Illuminate*, observes:

An example of the fallacy of the efficacy metric is seen in the comparison of a PAR30 halogen lamp and compact fluorescent lamp (CFL). A 50W PAR30 10° halogen spot produces 770 lumens, the same as a 15W triple-tube CFL. Based solely on efficacy, the 51.6 lumens/W CFL is superior to the 15.4 lumens/W halogen spot. However, at 12 ft., the PAR30 lamp will produce 48 fc, while the CFL produces less than 1. In a typical downlight, the PAR30 10° spot will suffer only a slight trimming of its perimeter beam pattern, while maintaining the 48 fc on the target; the CFL attains only 4 fc, and a beam spread of 98°, now at a luminaire efficacy of just 18 lumens/W. Similarly, a 61 lumens/W LED downlight delivers only 1.8 fc at a distance of 12 ft., with a 148° beam spread. Efficacy alone, without considering the design target, is a worthless metric. In this example, unrestrained efficacy metrics can favor products with poor optical distribution shaping, delivering high-angle brightness and limited glare control.

Figure 5-11 (left). One of the hottest HID markets is low-wattage ceramic metal halide track lights replacing incandescent and halogen sources. Due to low cost of miniaturization, electronic HID systems continue to get smaller towards a goal of making the ballast "disappear," driving smaller luminaire designs, reducing ceiling clutter and opening new track lighting and other applications for low-wattage ceramic metal halide (CMH) systems as an alternative to incandescent and halogen sources. With extremely components, CMH luminaire designs are approaching the form factor and size of low-voltage MR16 halogen systems. For example, at LIGHTFAIR 2010, Sylvania introduced a 15W T4 CMH lamp with a mini ballast, suitable for applications where a 20W system might be considered too bright. GE introduced three Micro electronic HID ballasts, a 20W and two 39W models, with one of the 39W models being the same size as the very compact 20W model, opening up new applications in this larger track lighting segment. Meanwhile, smaller-wattage CMH are getting better at rendering saturated reds with high R9 factors. The color rendering index is based on an average of the R1 to R8 palette; R9 through R14 are saturated colors commonly found in retail environments, and R9 is the saturated red color, particularly important to rendering colors as red is often found mixed into process colors. Lamps with high R9 values produce the most vivid colors. Halogen sets the bar with an R9 of 100. Many new low-wattage ceramic metal halide lamps are claiming high R9 values, rendering reds well, ideal for display lighting in food, flower, clothing and other applications. Photo courtesy of Lightolier.

An important lesson here is that illumination is more important than the device that delivers it. Good lighting design begins with lighting goals, not equipment. We must determine our illumination needs and then work backwards to what devices—lamps and luminaires—will satisfy these needs. Avoid focusing on lamps and luminaires as devices; instead, focus on the surfaces that these devices will be used to illuminate.

How Well Does The Light Source Render Colors?

Lamps produce light that has a color appearance, which can affect the color of objects when the light strikes them, and affects the color appearance of the lamp itself.

In review, the *color temperature* of a light source, expressed in kelvins (K), indicates the color appearance of the light source itself and the light it emits. Light sources are generally classified as "cool" (>4000K), which appear bluish-white; "neutral" (3000K-4000K), which appear white; or "warm" (<3000K), which appear orangish-white. Warm light sources are more heavily laden with red and orange wavelengths, bringing out some flesh tones and richer content in objects that have warmer colors (high- and low-pressure sodium being notable exceptions); a typical household incandescent lamp is a very warm light. Cool light sources are more heavily laden with blue and green wavelengths, enriching the visible color content of blue and green objects; daylight is a very cool light.

Color temperature describes the "whiteness," "bluishness," etc. of a light source, its visual warmth or coolness. However, it does not define how natural the color of objects will appear when lighted by the source. Two light sources can have the same color temperature, but render colors differently. The *color rendering index (CRI)*, a scale with a maximum rating of 100, offers a separate metric to address this.

For most common color temperatures, CRI uses daylight lamp as the reference for color rendering (100 CRI). In many applications, the higher the CRI the better, with 80-100 being optimal for rendering colors more "naturally"—that is, how most people would expect them to appear. Generally, the higher the CRI, the better. For a valid comparison of CRI ratings of two light sources, however, they must have the same color temperature.

For more information about color quality, see **Chapter 2**.

In the Spotlight: High Ballast Factor Ballasts

All fluorescent ballasts have a ballast factor (BF) rating that describes the percentage of light output from connected fluorescent lamps when the lamps and ballasts operate together as a system. Therefore, a lighting system consisting of a ballast with a standard BF of 0.88 connected to two 2,850-lumen fluorescent lamps will produce 2,850 x 2 = 5,700 x 0.88 = 5,016 lumens of light output.

Most ballasts are classified as low (0.74-0.78, typically 0.77 or 0.78) or standard (0.85-0.90, usually 0.87 or 0.88) BF ballasts. In the past 10 years, high BF ballasts (>1.0 BF, typically 1.15-1.18) have gained popularity as a means of increasing the light output of fluorescent lighting systems—most often 4-ft. and 8-ft. T8 systems—typically by 30-35 percent. (Some high-BF ballasts are also available for T5 and T12 lamps.) High BF ballasts produce more light and use more energy than standard BF ballasts, and operate at about the same efficacy. Most high-BF ballasts are instant-start, but pro- grammed-start high-BF ballasts have entered the marketplace for use in oc- cupancy sensor applications.

The primary application for high-BF ballasts are fluorescent hi-bay luminaires in new construction and retrofit projects as an alternative or replacement for standard metal halide luminaires. Replacing standard metal halide with high-bay fluorescent luminaires can result in energy savings approaching 50 percent, making this a hot retrofit market. High-BF ballasts enable high-lumen T8 lamped luminaires to be installed at higher mounting heights, making them a competitive alternative to T5HO hi-bay luminaires.

Be sure to operate the ballast only with lamp types and the designated number of lamps, avoiding mismatching the ballast with 2- and 3-ft. lamps, or taking a ballast rated for 3-4 lamps and operating 2 or 3 lamps instead. These conditions can result in overdriving the lamps, which may result in short lamp life. The ballast should incorporate active current regulation so that it does not overdrive its lamps if one lamp is removed from the circuit or fails.

Figure 5-12. The primary application for high-BF ballasts are high-lumen fluorescent T8 luminaires used to replace standard metal halide in high-bay applications such as this school gymnasium. Photo courtesy of Universal Lighting Technologies.

EVALUATING LUMINAIRES

A luminaire, commonly called a light fixture or lighting fixture in the field, is a complete lighting unit that produces and distributes light. It contains the light source, a ballast if the lamp is fluorescent or HID, components designed to diffuse or distribute the light in a controlled pattern, components to protect and position the lamp(s), and connection to the power source.

Luminaires may be categorized according to light source (e.g., fluorescent), application (e.g., indoor), function (e.g., downlight), layout/location (e.g., task), light distribution (e.g., asymmetric) or mounting method (e.g., recessed). These are not universal categorizations; luminaire manufacturers classify their products different ways.

Luminaire selection may be based on appearance, performance, mounting and maintainability. Luminaires are available in a variety of sizes, shapes and finishes. Because luminaires are often readily visible, aesthetics may be an important selection factor in some applications. The many selection criteria for a luminaire include:

- physical characteristics of the space such as room size, dimensions and materials, layout of furniture and obstructions such as partitions, room and object surface reflectances, structural obstructions such as beams;
- distribution pattern;
- glare control;
- luminaire efficiency (fraction of lamp lumen output transmitted out of the luminaire);
- appearance: importance of the luminaire being prominent (mostly decorative), neutral or integrating with the architecture;
- normal operating conditions, such as cleanliness of area during operation, maintenance schedule;
- special concerns such as safety, security, resistance to vandalism or hazardous or wet conditions;
- electrical, physical and operating characteristics of the light source selected, ballasts and controls;
- thermal and acoustical characteristics;
- finish;
- size;
- mechanical construction;

- accessibility of interior components for maintenance;
- availability of daylight;
- compliance with lighting power allowance set in applicable energy code; and
- reliability, availability and cost.

Based on the number of lamps in each luminaire, we determine the number of luminaires required to meet all design needs and then space them according to their distribution patterns.

Luminaire Efficiency

Today's focus on saving energy is accelerating demand for luminaires designed to achieve a high level of energy efficiency. Luminaires can be evaluated based on luminaire luminous intensity distribution or candlepower (candelas), total input watts (W), efficiency (fraction of lamp lumens that exit the luminaire), luminaire efficacy rating (lumens/W), coefficient of utilization (CU) and comparative yearly energy cost of light ($/1000 lumens).

While these metrics provide valuable tools for comparing the efficiency of luminaires, it is important to remember that efficiency is only part of the story of a lighting product and should be considered along with how the luminaire distributes the light and at what intensity. This will result in selection of luminaires that are both efficient and likely to achieve good visual comfort.

Luminaire efficiency is the ratio of light output emitted by the luminaire to the light output emitted by its lamps. Another way of looking at it: Luminaire efficiency is the percentage of light output produced by the lamps that are in turn emitted by the luminaire.

Not all light produced by the lamps will exit the luminaire; some will remain trapped inside and dissipate as heat. The luminaire's physical characteristics will affect how much light will exit and how much will be directed at the workplane.

Luminaire efficiency is important because while you can have a very efficient lamp-ballast system, if the luminaire itself is not efficient at delivering lumens, then the lighting system overall is not either. Factors that affect the efficiency of the luminaire include its shape, the reflectance of its materials, how many lamps are inside the luminaire (and how close they are to each other), and whether shielding material such as a lens or louver is used to soften or scatter the light.

While a high level of luminaire efficiency should be valued, over-emphasizing it can lead to poor lighting quality and angry users. After all, a bare lamp offers 100 percent efficiency, but is hardly a good choice. In reality, the most "efficient" luminaires are often candidates for direct glare, particularly unshielded luminaires with direct distribution at lower mounting heights typically found in offices, classrooms and similar applications. In such cases, light may exit the luminaire very efficiently, but the luminaire itself is a "glare bomb," and users may resort to wearing baseball caps.

Luminaire efficacy describes the efficacy of the entire luminaire, including the light source, ballast and luminaire losses. The *Luminaire Efficacy Rating (LER)* provides a metric for comparing the relative energy efficiency of fluorescent luminaires. Initiated in response to the Energy Policy Act of 1992, LER offers a voluntary rating standard for several categories of commercial and industrial fluorescent luminaires such as 2x4 recessed lensed and louvered luminaires, plastic wraparounds and striplights (see NEMA LE 5-2001 for more information).

LER is expressed:

$$LER = \frac{[\text{Luminaire Efficiency (EFF)} \times \text{Total Rated Lamp Lumens (TTL)} \times \text{Ballast Factor (BF)}]}{[\text{Luminaire Watts Input}]}$$

Some manufacturers publish LER in their products' photometric reports and specification sheets. Even without it, designers can easily calculate LER themselves as the information required by the above formula should be generally available for the product.

The *coefficient of utilization (CU)* metric allows us to look at luminaire efficiency in the context of the actual application. Since all room surfaces are potential reflectors of light, the room itself acts an extension of the lighting system. A given luminaire may emit some of its light directly at the workplane and some at a nearby wall. The wall absorbs some of the light and reflects the rest, some of which in turn reaches the workplane.

CU therefore allows us to compare luminaire efficiencies in a given environment. It shows the percentage of light output produced by the lamps that reaches the workplane after light is lost due to the luminaire's efficiency at transmitting light, the room proportions, and the ability of room surfaces to reflect light.

Luminaire manufacturers generally provide CU tables for their

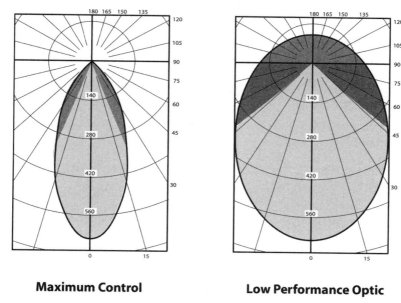

Maximum Control   **Low Performance Optic**

Figure 5-13. Efficacy is not the entire story of a luminaire. In the above drawing, the left luminaire operates at an efficacy of 28 lumens/W, while the right luminaire generates about 50 percent more light output for about 33 percent more wattage, resulting in a 14 percent higher efficacy. The right luminaire accomplishes this gain, however, through a lack of control of glare. Drawing courtesy of Kevin Willmorth.

products in photometric reports. As Average Maintained Light Levels (fc) = (Lumens x CU x Light Loss factors) ÷ Area (sq. ft.), CU can have a big impact on the capacity needs for a given lighting project and hence both its capital and operating costs. CU shows how changing room finishes can affect light levels.

The *Comparative Yearly Energy Cost of Light* is another luminaire comparison metric created in NEMA LE 5-2001 in response to the Energy Policy Act of 1992. It is expressed as a $/1000 lumens value based on the below formula:

$$\text{Energy Cost} = (K/LER) \times 1000 \text{ Lumens}$$

Where K = $0.24/W [(3,000 average operating hours per year x $ 0.08/kWh average energy cost) ÷ 1000]

Specifiers should be prepared to make adjustments as needed to tailor the formula to their project. The operating time averages to about 8 hours per day and can be adapted easily. The $0.08 per kWh cost is outdated as a national average and can also be adapted. According to DOE, the national average cost per kWh of electric energy was $0.1021 in 2009 for commercial buildings, increasing K to $0.31/W, and $0.0684 for industrial buildings, reducing K to $0.21/W. For the latest national averages and specific regional and even more specific state averages, visit http://www.eia.doe.gov/cneaf/electricity/epm/table5_6_a.html.

In the Spotlight: Protecting Your Design

Selection of the right luminaire and controls takes considerable time and effort, and is integral to the lighting design. Cost overruns, value engineering, unclear or weakly worded specifications, and conflicts between lighting and other buidling systems can result in substitutions and other design decisions being made without the lighting designer's involvement. Lighting designers should protect the integrity of their specifications and design vision through clear communication, defined quality standards and strong involvement with the owner and other design professionals during the construction process.

Luminaire Distribution

A luminaire's photometric characteristics are of paramount concern to the lighting designer, as they describe the direction and intensity of the luminaire's light output while also enabling calculation of luminaire efficiency and coefficient of utilization.

A luminaire's light distribution is determined by lamp type, the location of lamps within the luminaire, and by its optical system (i.e., reflectors, lenses, louvers, etc.). The particular characteristics of a given luminaire make it suitable for certain applications and not others. For example, the photometric performance of a fluorescent luminaire designed for an office is different than one designed as an alternative to HID sources in a high-bay warehouse application.

Design matters with luminaires. A poorly designed optical system may waste lumens and reduce efficiency while potentially creating direct glare and directly impacting visual comfort. Taking a T12 luminaire and adding T8 lamps and ballasts to it in the field during a retrofit will impact photometric performance. And a manufacturer taking a T12 luminaire model and replacing the T12 lamps with T8 or T5 lamps may also negatively impact performance.

As mentioned earlier in this chapter, luminaires are often classified by location relative to the workplane: general lighting, localized general lighting, local (supplementary) lighting and task/ambient lighting. They are also classified relative to the way they emit light: indirect, semi-indirect, direct-indirect, general diffuse, semi-direct and direct (see **Table 5-6**).

Direct luminaires distribute all or almost all of their light output in a downward direction that is highly concentrated or widely spread, depending on the optical system. Direct lighting is optically very efficient. In downlighting applications, strong direct lighting can create pronounced shadows and a tense atmosphere based on luminance contrasts. If the luminaire is located above and slightly in front of the task/occupant, overhead glare may result, which should be avoided in applications where stationary users perform critical tasks in a "head up" position. In offices and retail stores, recessed direct luminaires in offices have been criticized for producing strong scalloping on nearby walls, creating an atmosphere popularly termed as "the cave effect."

Indirect luminaires distribute 90-100 percent of their output in an upward direction towards the ceiling and upper walls, where it is reflected to all parts of the space. Totally indirect lighting also presents significant advantages and disadvantages as a design approach. On the plus side, indirect lighting scatters light in many directions, diffusing light distribution, which can aid visual comfort and facial recognition while eliminating shadows that are distracting and reduce visibility. However, using indirect lighting only can make a space appear flat and empty of highlights

Table 5-6. System for classifying luminaires according to how they distribute light. Note, however, that not all luminaire manufacturers follow this system strictly, and instead may define their luminaires more generally in their descriptions. Adapted from CIE.

Classification	% Uplight	% Downlight
Direct	0-10%	90-100%
Semi-Direct	10-40%	60-90%
General Diffuse	40-60%	40-60%
Semi-Indirect	60-90%	10-40%
Indirect	90-100%	0-10%

and shadows that provide spatial definition.

Indirect lighting turns room surfaces such as walls and ceilings into area sources and as such the performance of the system is more strongly related to the reflectances of these surfaces (see **Figure 5-14**). Durable, matte (or satin) room surface finishes with a high reflectance, regularly cleaned or repainted, are recommended; glossy finishes, which can reveal reflected images of light sources, are best avoided. When suspended luminaires are used, they should be mounted at a distance from the ceiling that is sufficient to provide a reasonably uniform brightness across the ceiling. Overall, the ceiling should not become a source of glare or unwanted reflections on computer screens.

Figure 5-14. Indirect lighting. Photo courtesy of Peerless Lighting.

Semi-indirect luminaires distribute 60-90 percent of their output in an upward direction and 10-40 percent in a downward direction. This type of system also uses the ceiling as an area light source and therefore the same considerations for indirect also apply to semi-indirect. *Direct-indirect* luminaires distribute a roughly equal amount of light up and down (40-60 percent) (see **Figure 5-15**). *Semi-direct* luminaires distribute 60-90 percent of the light in a downward direction while 10-40 percent is directed upward to generate desirable interreflections and reduce contrast between the luminaire and the ceiling. Some manufacturers use different terminology, such as "direct/indirect" to designate a luminaire with an up/down distribution that has a stronger downlight component, and "indirect/direct" to designate such a luminaire with a stronger uplight component.

Combining up and down distributions in the lighting design can combine the advantages of each approach while mitigating the tradeoffs. The direct component provides modeling definition and some light on the task plane while the indirect component provides diffuse ambient lighting.

Figure 5-15. Direct-indirect luminaire. Photo courtesy of Philips Lightolier.

In the Spotlight: Addressing Light Pollution

Light pollution obscures night-time viewing of stars, interferes with astronomical observations and wildlife, wastes energy and is a source of conflict between property owners. According to the International Dark Sky Association (IDA), an estimated 30 percent of energy used for outdoor lighting ends up flying into outer space—some $1.5 billion per year—while an estimated 100 million birds die each year from flying into buildings and other manmade structures because of light pollution.

Light pollution is a broad term encompassing three issues: sky glow (light emitted up into the sky, which can obscure night-time viewing of stars and interfere with astronomical observations), light trespass (light emitted into neighboring property) and glare (brightness that impairs or disables vision). These issues have united astronomers, environmentalists and communities to adopt legislation in a large number of municipalities across the United States, and resulted in a draft model outdoor lighting code (Model Lighting Ordinance, or MLO) produced by IES and IDA.

Figure 5-16. Good outdoor lighting can prevent light trespass by lighting only what needs to be lighted and using optics that prevent spill light onto neighboring properties. Photo courtesy of Architectural Area Lighting.

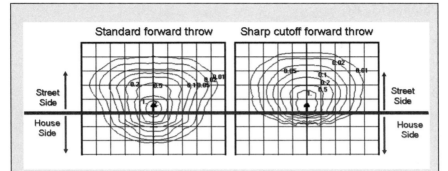

Figure 5-17. Type IV distribution (left) limits but does not sharply restrict backlight that can fall onto neighboring property. Type IV distribution with a sharp cutoff optic (right) produces forward throw distribution but sharply cuts off the backlight that can spill onto adjoining property. Graphic courtesy of Lithonia Lighting Outdoor Group.

In terms of design, the main issue is to avoid glare and place light only where it is needed. In equipment selection, this means choosing luminaires with low-wattage sources that produce no uplight and have good backlighting control. For area lighting in a parking lot, a 150-250W fully shielded shoebox-type luminaire can be a good solution. For lighting near the property perimeter, a sharp-cutoff, forward-throw optic will minimize light trespass.

The same goes for decorative and landscape lighting: Minimize the uplight. Decorative post-top luminaires are available that are both aesthetically pleasing and shield uplight with opaque sides or include an internal reflector to minimize uplight and glare. When lighting landscape or facades, use low-wattage sources and aim light in a downward direction wherever possible.

Indoor lighting can also do its part to reduce light pollution by minimizing light emission through windows during the night hours of 11:00 PM to 5:00 AM. Light should not be allowed to exit windows at all or lighting power should be reduced by at least 50 percent. Automatic window shades and automatic lighting shutoff controls can be used to accomplish this.

To evaluate outdoor lighting products, consider their BUG ratings. Developed jointly by the IES and the IDA, BUG goes beyond the old cutoff system to address backlight (B), a component in light trespass; uplight (U), which contributes to skyglow; and glare (G), which can be irritating and even disabling to vision.

Photometric Reports

In a perfect world, a lighting manufacturer would respond to interest in one of their products by assuming the cost of installing samples in an exact mockup of the actual space being designed, hiring people to work there for a while, and then conducting a post-occupancy survey on their satisfaction with their lighting.

In the real world, we have photometric reports. These reports are commonly available for specification-grade lighting products and can be found on the catalog sheet. What a report says about a luminaire can be used to predict how it is likely to perform in a given application, and thereby help us choose the right luminaire for the given application. Specifically, we can determine how the light is distributed, how efficiently it is distributed, and how likely it is to produce glare or unwanted patterns.

What are the basics that we need to know so that we can properly read and interpret photometric reports?

The most important items on the report are the candela chart and the candela distribution curve, more formally called the luminous intensity table and curve, which give us a picture of the luminaire's distributed light pattern. All the other items on the report, such as zonal lumen distribution, luminaire efficiency and luminaire spacing criteria are derived from the numbers in the candela chart's table. (Note that this discussion focuses on Type C photometry, which covers indoor general luminaires, and not Type B photometry, which is used for floodlighting and some other outdoor luminaires.)

Imagine that we are looking directly at the cross section (end) of a pendant-mounted linear direct/indirect luminaire in an open office. The luminaire has a vertical axis, an imaginary line running through its center from nadir at 0° (a point on the ground directly below the luminaire) up to 180° (a point on the ceiling directly above the luminaire), as shown in **Figure 5-18a**. It also has a horizontal axis that runs through its center from 90° to 270°. From our position looking at the end of the luminaire, we can take measurements of light intensity, measured in candelas (cd), at any angle of elevation from 0° to 180°, called *vertical angles*. In practice, these measurements are taken in manufacturer or independent testing facilities using a device called a goniophotometer.

We have now determined the light intensity values for a single *vertical plane* intersecting the cross section of the luminaire at its center. If the luminaire emits light in a perfectly symmetrical pattern in all directions, this would be enough to evaluate the luminaire's light distribution. But most luminaires—from 2 x 4 troffers to linear direct/indirect pendants to wall washers—do not.

This means we need to repeat the process of measuring light intensity at 0-180° from different positions around the luminaire to create more vertical planes and get the complete story. Instead of looking at the luminaire from the side, now we must look at it from the top and draw an imaginary line through its center as shown in **Figure 5-18b**. Typically, measurements are repeated at an angle parallel to the lamp axis (0°), 22.5°, 45°, 67.5° and perpendicular to the axis (90°), with 0°, 45° and 90° being the primary angles, as shown in **Figure 5-18c**. These are the *horizontal (also called azimuthal) angles*. Ninety degrees gets us a quarter of the way around the luminaire and is enough to give us a complete picture if the luminaire has a standard symmetrical geometric shape.

The result is a mapping of light intensities at different combinations of vertical and horizontal viewing angles. Visually, this three-dimensional pattern would look like an oddly shaped "bubble." Change the luminaire's reflector design, shielding and even just its lamp/ballast combination, and this bubble will morph into a new shape.

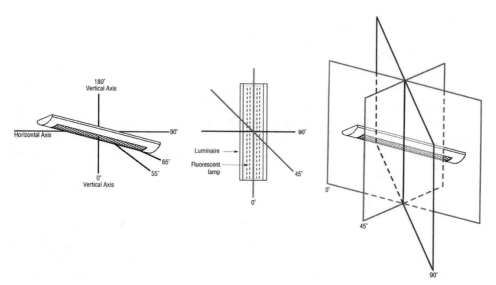

Figures 5-18a (left), 5-18b (middle), 5-18c (right). By measuring the intensities of light emission for a given luminaire at all major viewing angles surrounding it, we can create a three-dimensional pattern revealing its light distribution. Graphic courtesy of Finelite, Inc.

All of the above data are available in the photometric report's candela chart. The horizontal viewing angles (0°, 22.5°, 45°, 67.5°, 90°) are the column headings and the vertical angles (0-180° in increments) are the row headings. In **Figure 5-19**, we see an example for a direct/indirect luminaire. If we are facing the luminaire perpendicularly (the side of the luminaire, a 90° horizontal viewing angle) with our eyes at a 55° vertical viewing angle from the luminaire's 0° line (nadir), then the relative light intensity at that point is 234 cd. If we then circle the luminaire until we are facing its cross section (the end of the luminaire, a 0° horizontal viewing angle) at a 55° angle with the center of the luminaire (vertical viewing angle), then the relative light intensity at that point is 109 cd. At a 55° viewing angle, the luminaire emits more than twice the amount of light intensity in a 90° vertical plane than a 0° vertical plane.

The candela chart is important because it can be used for detailed analysis of a luminaire's light distribution and its impact on light levels and potential glare conditions using lighting design software. For this purpose, many manufacturers make the data available as downloadable electronic files on their websites. These files are typically based on a standard format created by the IES, which is why they are often called "IES files."

Note that the candela chart is generated based on a specific luminaire and lamp combination, so a three-lamp T8 fluorescent luminaire report will not apply to the two T8 lamp version of that same product, nor will it apply to a three-lamp T5 fluorescent model. Further, the actual ballast factor will have to be applied as a light loss factor. Similarly, if the light output of the specified lamps is different than those used in the photometric test on which the report is based, further adjustment will be necessary.

The *candela distribution curve* is a graphical representation of relative light intensity for a single vertical plane based on candela readings across the vertical viewing angles (0-180°) for a single horizontal viewing angle (see **Figure 5-19**). Since the distribution of light intensity varies based on the horizontal viewing angle, several patterns may be overlaid on top of each other; in this example, the pattern at a 90° horizontal viewing angle is shown as a solid dark line, a 45° angle as a lightly shaded line, and a 0° angle as a dashed line. The lines radiating from the center of the luminaire are the vertical viewing angles from 0° to 180°. And the concentric circles represent candlepower, or luminous intensity, with each progressive outward circle being a larger candela value.

While not as precise as the candela chart, the candela distribution curve can provide much of the same useful information and in an at-a-glance visual format. For example, looking again at **Figure 5-19**, suppose we would like to avoid a light intensity exceeding 300 cd at 55-90° vertical viewing angles because of glare concerns. Doing some simple eyeball estimating, candlepower is around 200 cd at 55° on a 90° vertical plane, 150-200 cd at 55° on a 45° vertical plane, and less than 100 cd at 55° on a 0° vertical plane.

Looking at the photometric report, probably the easiest thing to note is whether the luminaire is direct (the light is emitted below the horizontal axis), indirect (the light is emitted above the horizontal axis), or direct/indirect (a mix of the two, and to what degree). The luminaire in **Figure 5-19** emits 64 percent of its light output up and 36 percent of it down, so it is a direct-indirect luminaire (because more light is emitted up than down, it might be called an "indirect-direct" luminaire in the field).

We can also tell whether distribution is *symmetrical* (light output is emitted in a roughly equal pattern on both sides of the luminaire) or, as is common with cove lights and similar luminaires, *asymmetrical* (light output is restricted to one side or the other). If the luminaire has symmetrical distribution on both sides, only half of the drawing may be shown, as in **Figure 5-19**.

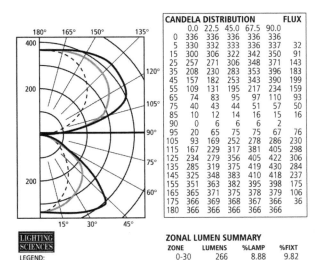

CANDELA DISTRIBUTION					FLUX
	0.0	22.5	45.0	67.5	90.0
0	336	336	336	336	336	
5	330	332	333	336	337	32
15	300	306	322	342	350	91
25	257	271	306	348	371	143
35	208	230	283	353	396	183
45	157	182	253	343	390	199
55	109	131	195	217	234	159
65	74	83	95	97	110	93
75	40	43	44	51	57	50
85	10	12	14	16	15	16
90	0	6	6	6	2	
95	20	65	75	75	67	76
105	93	169	252	278	286	230
115	167	229	317	381	405	298
125	234	279	356	405	422	306
135	285	319	375	419	430	284
145	325	348	383	410	418	237
155	351	363	382	395	398	175
165	365	371	375	378	379	106
175	366	369	368	367	366	36
180	366	366	366	366	366	

LEGEND:
0-deg. – – – – –
45-deg. ━━━━━
90-deg. ━━━━━

ZONAL LUMEN SUMMARY			
ZONE	LUMENS	%LAMP	%FIXT
0-30	266	8.88	9.82
0-40	448	14.97	16.55
0-60	807	26.90	29.74
0-90	965	32.19	35.60
40-90	516	17.23	19.05
60-90	158	5.29	5.85
90-180	1747	58.25	64.40
0-180	2713	90.44	100.00

Figure 5-19. Photometric report for a suspended 1-lamp T8 direct/indirect luminaire. Graphic courtesy of Finelite, Inc.

Additionally, we can tell whether the luminaire has a spot distribution (narrow pattern), narrow and medium flood (fuller pattern and a flatter bottom), or wide flood (wide pattern and possibly a "batwing" shape where peak distribution is on each side of the center instead of directly above or below the luminaire). We can tell whether the luminaire is likely to produce a smooth light pattern (smooth, rounded candela distribution curve) or "streaking" on walls or the floor (striations in the pattern). And we can tell whether the luminaire is likely to produce glare (a high concentration of direct light intensity is being emitted at a 55° vertical viewing angle in an office or above 60° in another application). An experienced eye can learn even more than that at a glance.

Other interesting data in the photometric report are derived from the light intensity measurements, such as zonal lumen summary and luminaire efficiency.

The *zonal lumen summary table* lists the luminaire's light output, in lumens, in specific zones and then summarizes for all light emitted down (0-90°), up (90-180°) and total (0-180°). These values are used to calculate the *luminaire efficiency*, the percentage of lamp light output (in lumens)

Table 5-7. Sample (illustrative) candela distribution curves for CIE luminaire classifications.

Classification	% Uplight	% Downlight	Typical Candela Distribution Curve
Direct	0-10%	90-100%	
Semi-Direct	10-40%	60-90%	
General Diffuse	40-60%	40-60%	
Semi-Indirect	60-90%	10-40%	
Indirect	90-100%	0-10%	

that exit the luminaire relative to the total lamp lumens that go into the luminaire. (Luminaire Efficiency is the Sum of All Zonal Lumens ÷ Nominal Lamp Light Output x 100). The luminaire portrayed in **Figure 5-19**, for example, has an efficiency of 90.4 percent. But while higher efficiency is generally better, we must consider where that light is going to see if the emitted light is actually useful. Unshielded fluorescent striplights can be as efficient as 95 percent, but again would be considered a "glare bomb" by office workers. There is often a tradeoff between luminaire efficiency and optical control: The more the luminaire works to deliver light where it is wanted and block it where it is not wanted, the lower its efficiency will be.

Initial cost, aesthetics, ability to provide target light levels, and lamp/ballast efficiency—these are all important considerations when choosing a luminaire. But they say nothing about how the luminaire will actually perform in the space—and what impact it will therefore have on the people who use the space for work or leisure. What we really need to know is how the light is distributed, how efficiently it is distributed, and how likely it is to produce glare or unwanted patterns.

It's all in the photometric report.

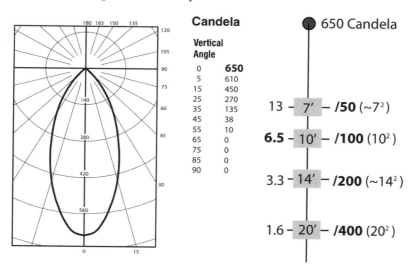

Figure 5-20. **Candela can be used to calculate light level at the center of the luminaire using the Inverse Square Law: candela ÷ distance² = footcandles. At 10 ft., for example, light level is simply calculated by dividing intensity by 100. Drawing courtesy of Kevin Willmorth.**

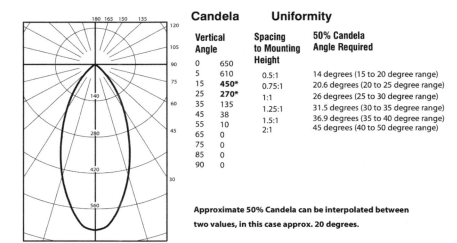

Candela

Vertical Angle	
0	650
5	610
15	**450***
25	**270***
35	135
45	38
55	10
65	0
75	0
85	0
90	0

Uniformity

Spacing to Mounting Height	50% Candela Angle Required
0.5:1	14 degrees (15 to 20 degree range)
0.75:1	20.6 degrees (20 to 25 degree range)
1:1	26 degrees (25 to 30 degree range)
1.25:1	31.5 degrees (30 to 35 degree range)
1.5:1	36.9 degrees (35 to 40 degree range)
2:1	45 degrees (40 to 50 degree range)

Approximate 50% Candela can be interpolated between two values, in this case approx. 20 degrees.

Figure 5-21. Candela can be used to calculate rule-of-thumb luminaire spacing. For good uniformity, beam patterns should be overlapped at 50 percent of maximum candela, with a rule of thumb being to add 5° for every ¼ spacing starting from 0.5:1. For example, at a 1:1 ratio, the distance from the luminaire to the workplane is the same as the distance between the center of each luminaire. For the above luminaire, at 1:1, a beam pattern with 50 percent maximum candela will have to be 25-30°. At 1.5:1—representing two additional ¼ increments—the beam pattern will have to be 35-40° to achieve the same uniformity. Drawing courtesy of Kevin Willmorth.

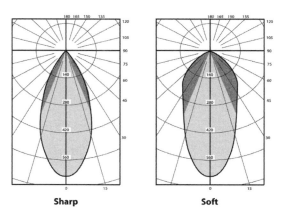

Sharp   **Soft**

Figure 5-22. In this candela distribution curve, the darker shaded areas identify the area above the main beam, or 50 percent distribution angle. The less area above this angle, the sharper the edge of the beam pattern will be. The left curve has very little intensity above the main beam and therefore the luminaire has a sharp-edged distribution. The right curve has more light distributed above the main beam and therefore has a softer-edged light pattern. Drawing courtesy of Kevin Willmorth.

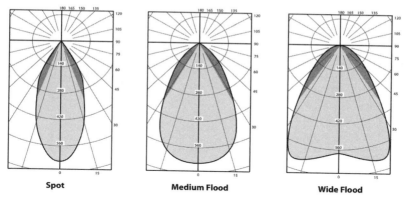

Spot   **Medium Flood**   **Wide Flood**

Figure 5-23. A brief comparison of candela distribution curves can quickly narrow a series of choices by indicating whether the luminaire produces a spot (narrow patterns), narrow or medium flood (fuller pattern and flatter bottom at the center), or wide flood (widest pattern and, in some cases, a batwing distribution, indicating that peak distribution straddles the center rather than being directly below the luminaire). Drawing courtesy of Kevin Willmorth.

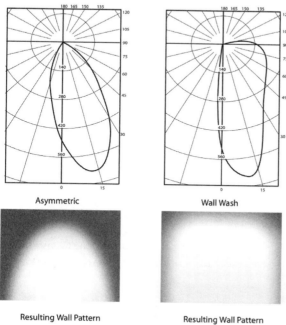

Asymmetric   Wall Wash

Resulting Wall Pattern   Resulting Wall Pattern

Figure 5-24. The luminaire on the left places some light on adjacent vertical surfaces, but the luminaire on the right is a wall washer, intended to illuminate the vertical surface uniformly from top to bottom—hence its particular distribution of intensity. While the wall washer would be preferable for washing a wall evenly across its surface, note the distribution at angles that may produce glare, and therefore care should be taken to avoid direct glare for occupants. Drawing courtesy of Kevin Willmorth.

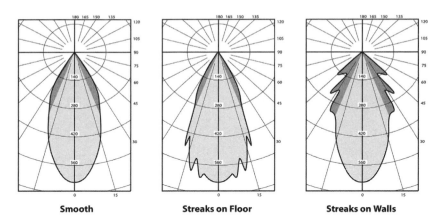

Figure 5-25. The candela distribution curve indicates the resulting pattern on task surfaces. Spikes and bumps indicate striations that may be mirrored on the task surface. When these features appear near the center of the pattern, their effects will show up on horizontal surfaces. When they appear at higher angles, as shown in the right candela distribution curve, they will show up on vertical surfaces. Drawing courtesy of Kevin Willmorth.

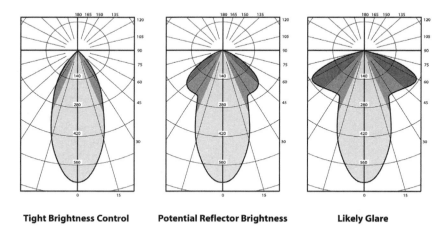

Figure 5-26. The candela distribution curve can be used to identify significant intensities above 60°, which are likely to produce glare. Drawing courtesy of Kevin Willmorth.

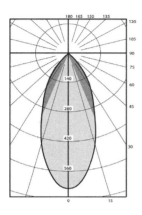

Maximum Control

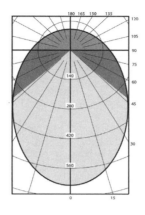

Low Performance Optic

Figure 5-27. The closer the beam pattern is to a round shape for a given luminaire, the less likely it is to have an effective optical system, as shown in a comparison of these candela distribution curves. Drawing courtesy of Kevin Willmorth.

In the Spotlight: High-Efficiency Recessed Lighting

The proliferation of computers in office buildings resulted in a shift in demand from lensed to parabolic troffers because parabolic luminaires use shielding at high angles to reduce glare on computer screens. Parabolics are now a staple in offices, schools and other commercial and institutional fluorescent applications. The problem with this approach is that by reducing shielding high-angle emission, they tend to produce scalloping on nearby walls, creating a dim atmosphere popularly known as the "cave effect."

Computers have changed over the years, however. Older screens are sensitive to luminaire brightness in directions approaching horizontal. Recent advances have virtually eliminated this sensitivity. For example, today's computer screens use a larger radius of curvature, improved screen brightness and anti-reflectance technologies. Flat screens, meanwhile, are becoming increasingly common. And software developers typically rely on positive contrast—dark characters on light backgrounds—to present information.

Enter a new type of fluorescent troffer that has gained some popularity. Possibly better known as "volumetric lighting," the generic term popularized by Lithonia Lighting when it introduced RT5 in 2004, this luminaire type is also known by other terms, such as "diffuse lighting." Examples include Lithonia's RT5 and RT8; Day-Brite's Attune and SofTrace; HE Williams' HET-T8 and HET-T5; Columbia's EPOC, Zero Plenum Troffer and Energy Max Intersect; and Cooper's Accord, Ovation, 2HP and Class R products.

Figure 5-28. Volumetric lighting can mitigate the cave effect common with parabolics, making spaces appear brighter, while reducing energy consumption. Lighting quality is improved because illumination is placed on the high vertical surfaces in the space, as shown here in this comparison between a room lighted with 3-lamp parabolics (left) and 2-lamp high-efficiency premium troffers (right). Photos courtesy of Day-Brite Lighting.

These products feature photometric distribution including a small amount of light in the 60- to 90-degree zone, smoothly fading as the angle gets closer to 90 degrees (see **Figure 5-29** and **Figure 5-30**). A typical product consists of a central luminous body with a shallow reflective surround, presenting an enhanced architectural aesthetic that blends well with upscale and other contemporary interiors. The light source, which may be T5, T5HO, T8 or compact fluorescent, is concealed behind diffuse shielding. Light output can be modulated by using lamps with different levels of light output and ballasts with different ballast factors. Downward distribution typical of direct lighting is combined with light emitted in directions approaching horizontal.

This combination results in light being distributed uniformly on high vertical surfaces in the space, mitigating the cave effect and improving visual comfort (see **Figure 5-28**). Manufacturers also claim that these luminaires have a good energy story, able to replace three-lamp parabolic and lensed T12 troffers in older overlighted spaces with energy savings as high as 50 percent.

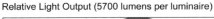

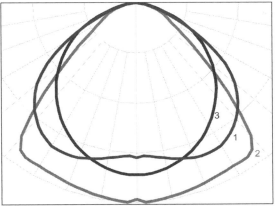

Relative Light Output (5700 lumens per luminaire)

1 Day-Brite Attune Controlled Vertical Luminance product
2 Typical commodity (2) T8 A12 Prismatic Troffer
3 Typical commodity shallow louver (2) T8 12 Cell Parabolic Troffer

Figure 5-29. Light distribution for a high-efficiency luminaire compared to the prismatic lensed and parabolic louvered luminaires with the same light output. Note the higher light emission in the vertical angles, controlled to avoid glare. Drawing courtesy of Kevin Willmorth.

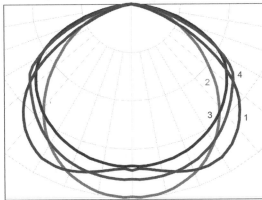

Relative Light Output (Normaized to 5700 lumens per luminaire)

1 Day-Brite Attune Luminaire
2 Lithonia RT5 T5 Luminaire
3 Metalux Accord Luminaire
4 Columbia Epoc Luminaire

Figure 5-30. Comparison of distributed light intensity among high-efficiency recessed products from different manufacturers, normalized to relative light output of 5,700 lumens. Despite some differences, all four will produce similar results. Drawing courtesy of Kevin Willmorth.

EVALUATING LIGHTING CONTROLS

Lighting controls play a critical role in electric lighting systems, providing the function of turning the lights on and off using a switch and/or adjusting light output up and down using a dimmer. Lighting controls provide the right amount of light where it's needed and when it's needed, supporting both energy management and visual needs.

In recent decades, technological development has increasingly automated these functions and allowed integration of devices into larger, more flexible systems. Today's lighting controls can provide significant benefits, including lower energy costs, improved worker satisfaction and greater flexibility.

The largest and fastest-growing application for lighting controls is energy management. According to the New Buildings Institute, lighting controls can reduce lighting energy consumption by 50 percent in existing buildings and by at least 35 percent in new construction, while reducing peak demand charges.

As a reliable energy-saving strategy, automated lighting controls are now mandated by commercial energy codes, supported by many utility rebate programs, and positioned as an integral component in sustainability initiatives such as LEED.

This section describes major advanced lighting control options and closes with a discussion about the most economical strategies that incorporate lighting controls into lighting upgrades. For more information, see the author's companion volume, *The Lighting Controls Handbook*, also published by The Fairmont Press, and AboutLightingControls.org, the Lighting Controls Association's website, which offers whitepapers, monthly newsletter and online education courses about lighting control technology and application.

Automatic Shutoff Controls

Most energy codes require some type of automatic shutoff, either from an intelligent control panel, occupancy sensor or building automation system (BAS). While required in new construction, lighting automation is also highly suitable for retrofits.

An intelligent relay panel is a low-voltage control system that turns the lights off at a preset time of day, producing 5-15 percent total building energy savings (see **Figure 4-25**). Its operation is simple: An integral programmable astronomical time-clock sends a low-voltage signal to relays, which break power to the load. The relays can be connected to other controls emitting a compatible low-voltage pulse, such as a photosensor or BAS.

Intelligent relay panels provide the backbone of an energy-saving lighting control system for switching and/or dimming, supplemented by manual switches, occupancy sensors, photosensors, etc. They are ideal for controlling lighting in large areas where large numbers of people work a predictable schedule. In spaces where scheduling is used, occupants should be given a local override to keep the lights on in the immediate area.

While programmable time-clock-based controls turn lights on and off based on a schedule, occupancy sensors turn them on and off based on whether the monitored area is occupied, producing 35-45 percent energy savings.

Occupancy sensors are advantageous in that they provide economical distributed control—i.e., ability to control a localized space—which can result in higher energy savings; additionally, local override is usually part of the sensor's functionality. A good rule of thumb is to use occupancy sensors in private spaces where there is an unpredictable rate of occupancy and across spaces where more granular control is desired—e.g., private offices, copy rooms, restrooms, etc.

Consider special features when selecting occupancy sensors, such as dual-technology and self-calibration for maximum reliability, manual-on for maximum savings, integral dimmer, and integral photosensor that keeps the lights off if there is enough daylight. To maximize savings, don't forget to control workstation task lighting with occupancy sensors as well as the general lighting.

Figure 5-31. Programmable intelligent relay panels can provide the backbone of an energy-saving lighting control system for switching and/or dimming, supplemented by manual switches, occupancy sensors, photosensors, etc. Photo courtesy of Schneider Electric/Square D.

In the Spotlight: Digital Revolution

Digital lighting control systems tie together intelligent control devices in a network enabling cost-effective integration of multiple control strategies such as scheduling, occupancy sensing, daylight harvesting, personal dimming control and demand response.

Other benefits include individually addressable luminaires for the ultimate in flexibility, simplified wiring, plug-and-play installation, elimination of centrally located equipment such as relay panels, software-based remote configuration and commissioning, ability to make relatively easy changes and expansions, and the potential for real-time energy reporting.

Solutions vary by manufacturer, and many have gone beyond the DALI protocol to offer proprietary solutions that either use relays or dimming ballasts as the control platform. The best-in-class solutions offer the ultimate in lighting control flexibility and energy savings.

Daylight Harvesting

Daylight harvesting occurs when a photosensor measures task illumination and signals a control to adjust light output to maintain the desired task light level. Daylight harvesting is required for certain spaces by California's Title 24 energy code but few other codes; in coming years, however, this is expected to change.

A basic daylight harvesting system starts with a photosensor that measures task light levels in areas receiving daylight (spaces near windows and skylights) and signals a controller to adjust the lighting system's light output when a target is reached. The control mechanism can be a dimming ballast that continuously tunes light output to maintain the set light level, a stepped-dimming system that changes light levels between several points but with smooth transitions, or a switching control that turns the lights on or off. Because an effective daylight harvesting control system saves energy while being virtually unnoticed by occupants, dimming is recommended for spaces where occupants perform stationary or critical tasks, where daylight provides only a portion of the required light level or daylight contribution is variable, and where luminaires must be mounted in normal field of view or lamps can be visible.

As with other controls, energy savings will depend on the application, but daylight harvesting controls can produce 35-60 percent energy savings, according to the New Buildings Institute. Daylight harvesting is what makes daylighting a sustainable strategy.

Architectural Dimming

The project may be driven by business benefits other than energy savings by providing increased performance and flexibility afforded by architectural dimming.

For example, the lighting can be adaptable to multiple uses of a space, such as conference room, gymnasium, ballroom or house of worship, or to evolving needs resulting from employee churn and strategies such as hoteling/hot-desking (in which workers do not have permanent seat assignments and share workspaces). Flexibility in selection of light levels can also be used for mood-setting in restaurants and similar applications, and to enhance aesthetics and image.

These business benefits are often more difficult to quantify than energy savings, but tangibly contribute to the bottom line.

When specifying architectural dimming for an application with multiple switch-legs creating a layering effect with the lighting (such as a house of worship), consider preset (integrated) dimming, which consolidates multiple dimmers into a single keypad-based dimming control system. Each keypad, or control station, can be programmed to create repeatable lighting "scenes" that can be recalled at the push of a button. These controls are relatively easy to build networks with and integrate into larger control stations.

Personal Control

Studies have demonstrated that people respond very differently to their environment regarding lighting and temperature. For lighting, this is expressed as varying preferences for lighting conditions. Conventional lighting systems, however, do address these variable needs, being one-size-fits-all systems designed to satisfy a majority, but not all, occupants.

Personal control allows users to set light levels to their personal preference and thereby gain more control over their lighting conditions. Personal control strategies include switching and dimming, and can include multiple layers of lighting, such as task lighting. Switching strategies include bi-level or multi-level switching. Dimming strategies include bi-level and continuous dimming of lighting layers down to the granular level of individual luminaire control, with the user controlling his or her own lighting using a PC interface, desktop control or handheld remote.

Personal control has been the subject of intensive research and has been demonstrated to increase productivity among office workers—specifically in the metrics of satisfaction, motivation and vigilance—while generating 10+ percent energy savings. Both dimming and bi-level or multi-level switching can be used to comply with the bi-level switching requirement in energy codes based on the IECC.

If designing for control zones as small as individual luminaires, consider a digital lighting system utilizing the Digital Addressable Lighting Interface (DALI) or appropriate proprietary protocol. Digital lighting networks enable facility operators to set up control zones as small as a luminaire and reconfigure them without rewiring.

In the Spotlight: Combining Control Strategies

Lighting control strategies can work well together—in some cases even within the same luminaire. For example, National Research Council Canada (NRC) – Institute for Research in Construction conducted a year-long study of luminaires integrating occupancy sensing, daylight dimming and individual user dimming control. The controls produced 42-47 percent energy savings, while surveys of participating users showed a correlation between the advanced lighting system and higher satisfaction.

The researchers credit the system's success in part to the presence of a dedicated local employee ensuring that the system was well-maintained, particularly with respect to ensuring the IT department re-enabled the individual dimming control for new occupants.

A Role for Control in Lighting Upgrades

According to the New Buildings Institute, advanced lighting controls can generate up to 50 percent lighting energy savings in existing buildings, which may justify the addition of lighting controls to relighting projects as well as lamp and ballast retrofits.

The biggest challenge is adding low-voltage control wiring to an existing space, generally limiting opportunities for installation of sophisticated control systems. As a result, the simplest upgrade options involve the least amount of rewiring or simply swapping out older ballasts and controls for new controls.

The first lighting control strategy to consider is automatic shutoff. It is considered the easiest, lowest-risk path to energy savings and is relatively simple to set up and commission. If LEED for existing buildings is used as a model path or actual requirement for the upgrade, this will be essential, as LEED requires that buildings meet the ASHRAE 90.1 energy standard as a prerequisite to gaining points for transcending it.

Start at the lighting panel. Are there large, open spaces in the building with predictable hours of operation? Are there public spaces where the lights must stay ON even when a space is unoccupied? If so, consider upgrading the existing lighting panelboard to an intelligent lighting control panel that offers programmable scheduling. Be sure to give local users override capability with a maximum 2- to 4-hour override.

Next, consider replacing the wall switch. Are there smaller, enclosed spaces in the building that are intermittently occupied during the day and are lighted with instant-ON light sources? If so, consider replacing toggle wall switches with occupancy sensors. If there is a clear line of sight between the switch and the primary task area, PIR sensors can present a cost-effective option. If greater sensitivity is needed for small levels of motion or if there are obstacles between the wall switch and the task, consider ultrasonic. For the ultimate in reliability, consider dual-technology sensors.

If the space is a private office already circuited for bi-level switching, consider replacing the manual switches with a manual-ON/auto-OFF occupancy sensor for the highest positive energy savings and some flexibility (See **Figure 5-32**). If the space requires an occupancy sensor be installed in a location other than at the wall switch, consider wireless occupancy sensors that run on batteries or ambient light in the space harvested using an integral solar cell. These sensors install anywhere within range of the receiver switch, which replaces the wall switch, and present

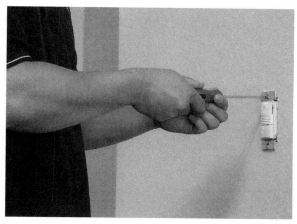

Figure 5-32. The easiest controls retrofit involves replacing components with the least amount of rewiring. While this often leads to occupancy sensors and lighting panelboard upgrades, new wireless controls and the falling cost of dimming ballasts are expanding the potential role for lighting control in building upgrades. Photo courtesy of WattStopper.

no wiring requirements, although wireless technology is presently a premium option. Similarly, wireless photosensors are also available.

If the upgrade involves replacing luminaires, consider integral controls. In a workstation-specific open office lighting layout, for example, direct/indirect luminaires can be installed that include an integral occupancy sensor and/or, if placed in a daylight zone, a photosensor and dimmable ballast, with the control wiring located inside the luminaire. If the space is a high-bay lighting application where metal halide is being replaced by fluorescent luminaires, consider luminaire-integrated or mounted line-voltage occupancy sensors, which can be an economical addition to a new fluorescent luminaire or separate add-on that is field installed (see **Figure 5-33**). Photosensors could be similarly added for control of luminaires mounted over spaces that receive ample daylight from skylights.

Light levels can be stepped using a single ballast called a step dimming or light level switching ballast. If the existing space is already circuited for bi-level switching, light level switching ballasts can be installed to ensure light levels are reduced uniformly, without a checkerboard pattern. These ballasts can operate without low-voltage wiring. Most products are programmed-start T8 ballasts, which may experience a loss of efficacy during light level reduction. Instant-start light level switching ballasts are available that offer proportional reductions in light output and input watts, although instant-start operation is not recommended by some manufacturers for applications with five or more switching cycles per day (see **Figure 5-34**). Other hi/lo switching opportunities include corridors that receive a lot of daylight (with a photosensor) and stairwells (with an occupancy sensor).

Continuous-dimming ballast costs have been falling for years, putting this control method within reach of many upgrade projects. Efficiency has also improved such that dimmable ballasts are available that are as efficacious as standard instant-start fixed-output ballasts. Look for the NEMA Premium label for the most efficient ballasts.

Some dimming ballasts are available that communicate with lighting controls using existing line-voltage wiring. Two-wire phase-control dimming ballasts use existing line-voltage lines for both power and communication and are suitable for any application where greater flexibility is desired, such as conference rooms, boardrooms and private offices. A dimming range of 100-5 percent is available for T8 lamps and CFLs, and 100-1 percent for T5HO lamps. The lighting is typically controlled via local controls accessible to occupants.

Figure 5-33. Acuity Brands Lighting, the parent of Lithonia Lighting, installed the company's I-Beam fluorescent high-bay luminaires with integral line-voltage occupancy sensors in one of its warehouses, generating significant energy cost savings. Photo courtesy of Lithonia Lighting.

Figure 5-34. Light level switching ballasts provide uniform light level reduction without a checkerboard pattern. Photo courtesy of Universal Lighting Technologies.

Line-voltage stepped dimming ("load shedding" or "demand response") ballasts may be combined with specialized energy management systems enabling a preset light level reduction, with a fade transition between light levels, in response to a variety of control inputs such as photosensors and schedules. The ballast may be combined with a signal transmitter that initiates load shedding in response to some type of demand response program. While demand response is still emerging as a trend, it will likely play a larger role in lighting in the future.

The ultimate control upgrade involves creating a fully realized lighting control system combining multiple strategies. In spaces where stationary tasks are performed, dimming will be preferable to switching while the space is occupied. If the ballasts will be replaced with dimmable ballasts, then multiple strategies should be enacted to make this installation more economical. When wiring a control system enacting multiple strategies around a dimmable ballast, one should consider a digital communication architecture, which eliminates multiple home runs and produces installation savings. If a digital architecture is chosen, one can consider creating a system out of DALI-compatible components, or specifying a proprietary system built around relays in distributed power packs and occupancy sensors, or digital dimming ballasts.

Finally, if the existing installation already includes automatic lighting controls that will be retained after the upgrade, ensure these controls are working properly by re-commissioning them as part of the project. The system may have been improperly designed, installed or commissioned when first put in place, or its operating parameters may have drifted out of sync with the space and how its lighting is used. Re-commissioning can therefore become a source of energy savings by itself.

The bottom line is that in most spaces, simple control strategies can be economically incorporated into lamp/ballast upgrades and luminaire replacement projects, accelerating energy savings and, in some cases, improving flexibility.

SOLID-STATE LIGHTING: LEDs

DOE predicts that solid-state lighting will achieve efficacies as high as 160 lumens/W and provide most of our general lighting needs within the next 20 years, thereby reducing the nation's energy costs by 6-7 percent.

The revolution will be illuminated, and it's already begun. This is an exciting time for solid-state lighting—an era where LED lighting

Figure 5-35. Bi-level lighting system controlled by a patented motion sensor, designed for sparsely occupied areas such as stairwells, corridors and conference rooms. When the space is occupied, the lights come on full; when it is unoccupied, the lights gradually dim to one of four preset light levels. Photo courtesy of LaMar Lighting.

is making big strides. Technology is still advancing and costs are coming down, and needed industry infrastructure, such as critical testing standards and mechanisms for quality recognition, is being delivered. Five years ago, LED general lighting was more concept than reality. Though 2007, LED lighting applications were limited to niches such as exit signs, accent lighting, decorative lighting and color LED applications. Now LED products are being introduced for a broad range of display, commercial, industrial and outdoor area lighting applications, even troffer and high-bay luminaire replacements. Everybody is asking for it, everybody is making it. The technology is moving so fast, in fact, that next-generation products are entering the market every six months. According to Strategies Unlimited, the global market for LED lighting will top $5 billion in 2012, a compound annual growth rate of 28 percent from 2008-2012.

LED products took top honors at 2009 LIGHTFAIR International in New York City with multiple recognition at the event's Innovation Awards, a sign that LEDs were starting to become a serious contender in the general lighting market. At 2010 LIGHTFAIR, it became a flood, with roughly 130 out of the 200 products entered into the Innovation Awards based on LED or another solid-state light source. That's two out of every three products. And an LED product—an LED lamp module—was named Product of the Year. And according to time-honored tradition that the show be nicknamed according to a dominant emerging product type displayed at a large number of booths, LIGHTFAIR once again became "LEDFAIR."

How are LEDs Different than Traditional Sources?

While LEDs need to be considered as just another light source that can get a lighting job done, the fact is that LEDs are a unique technology, even if it often comes in familiar luminaire packaging. In what critical ways are LEDs different than (and in some cases similar to) traditional light sources?

First, LEDs produce light using a non-traditional method. With conventional sources, electric energy is converted to light by heating a filament (incandescence) or passing current to excite a gas (fluorescence and high intensity discharge). With LEDs, current is passed through crystalline solids (which makes LEDs a "solid state" light source and explains why it is more formally called "solid state lighting") to produce visible light. This construction yields products that are far more robust, impervious to vibration and immune to extreme cold, with no mercury.

Figure 5-36. At Yum! Brands' first LEED Gold-certified restaurant, a KFC-Taco Bell built in Northampton, MA in 2008, lighting designer Derry Berrigan specified white LED lighting for 95 percent of the illumination needs, providing lighting that is aesthetically pleasing and visually comfortable while also being highly energy-efficient. As a result, the interior lighting achieves 46 percent energy savings compared to a system meeting the requirements of the ASHRAE 90.1-2007, while the exterior lighting achieves 77 percent energy savings. The design includes solutions such as Cree's LR4, LR6, LR6C and LR24 products; Insight Lighting's Tre'o Series; Finelite's Personal Lighting System (PLS); and BetaLED's Edge System. Photo courtesy of Derry Berrigan Lighting Design.

LED devices used in architectural luminaires are typically high-output packages that contain die (the LEDs themselves) and some form of encapsulated enclosure mounted on a base to facilitate mounting to circuit boards and heat sinks. Traditionally, luminaires and components are separate items built to widely accepted standards; items from different manufacturers (lamp, ballast, luminaire) are mixed and matched to create a finished product. In contrast, LED luminaires are usually integrated systems consisting of the LED device, optics, a heat sink, drive electronics and luminaire provided as a complete unit. The "lamp" is typically integral to the system and cannot be removed or replaced.

LED product performance is highly dependent on the environment in which these products operate; the only valid performance data for LED products, therefore, is based on total luminaire performance, not the base LED output. This is actually true of conventional sources as well, where luminaire efficiency often presents a significant loss of performance, but with LEDs, it is more critical.

Colored LEDs emit very narrow bands on the color spectrum, which makes this light source ideal for generating saturated colors. The combination of red, blue and green (RGB) LEDs can also be mixed to create white light, although of questionable quality. Most white-light LED products instead utilize phosphor-coated blue or ultraviolet LEDs to create white light output the same way that fluorescent lamps do, and with very similar results in both color quality and color selection.

Like traditional light sources, the color temperature (color appearance or tone of the light source and the light emitted) of LEDs can be specified as warm white (2800K-3000K) or cool white (4000+K), but with several differences. As with fluorescent sources, warm-white phosphor-coated LEDs typically have a lower efficacy (lumens of light output per watt of electrical input, or lumens/W) than cool-white phosphor-coated LEDs. Note that some LEDs can go as high as 6500K (daylight) or even higher; above 6500K is considered too cool for many architectural applications. In addition, it is likely that a given luminaire will have good LED-to-LED color consistency, but there may still be slight differences in color temperature between luminaires, which may or may not be noticeable (more noticeable when luminaires are very close to each other). One expert on LEDs explained that one can expect to see color variations between individual LED luminaires similar to that experienced with today's compact fluorescent lamps (making an interesting, larger point that we must be careful not to hold LEDs to a perfect performance standard for issues that plague many or even all lighting devices).

Regarding color rendering, a CRI rating of >80 is often recommended for commercial applications, which is now available in products using quality LEDs. In fact, CRI ratings as high as 92 are now available, and will continue to improve as the technology improves. CRI ratings of RGB (color) LED systems, however, can be as low as 13-32 CRI, an area of concern for applications where color rendering is important.

Note that in either case, there are arguments that CRI is an imperfect metric, particularly when it comes to LEDs, and particularly with RGB. As with traditional light sources, there is no substitute for seeing firsthand how a given light source renders skin tones and finishes; when in doubt, "test drive" the light source first if possible.

The light emitted by the LED contains no direct heat or ultraviolet light, unlike conventional sources. However, that does not mean that heat is not an issue. LEDs generate as much heat as any other source of similar wattage, with one large difference—all of the heat is emitted from the back of the device, and must be removed or light output, efficiency and useful life of the LEDs and related electronic devices will be severely degraded. This requires the LED luminaire to contain significant thermal management features. Because of this, products without apparent and significant components to extract heat from the LEDs should be avoided.

Like conventional sources, LED devices have a declining rate of lumen depreciation over their life. In fact, being an electronic device means that failure modes are quite different from standard lamps, which burn out long before their decline in light output becomes a significant issue. LEDs may last as long as 100,000 hours before failure, but at some point the amount of light being produced is so low that replacement is necessary to restore light levels. Because of this, service life of an LED product is rated at the time in hours of operation where the product is producing 70 percent lumen maintenance (30 percent loss of light output, about 15,000-50,000 hours for an LED device depending on whether the LEDs are overdriven and by how much) for general lighting applications and 50 percent lumen maintenance for decorative and other non-critical applications (up to 50,000-70,000 hours in well-designed products is possible).

The owner gets a double maintenance benefit from LEDs: long service life plus no spot replacement of failed lamps. The downside: The owner will have to remember, possibly as long as a decade-plus into the future, to change out the LED luminaires when they have reached their end of service life, which may not be apparent in any actual failures. This presents particular challenges in applications (like egress corridors) where codes demand minimum light levels be maintained.

LED Modules

One of the factors inhibiting development and acceptance of LED lighting is there are still no standards for connecting LED light engines and power supplies to luminaires as there are for conventional technologies such as fluorescent and HID. Luminaire manufacturers typically build products around a given light engine, resulting in many customized light engines and light sources that cannot be replaced since LED systems are typically hardwired into the luminaire, with no thought given to how to handle failures or upgrades.

If the light source were easily replaceable, then the user gains a durable luminaire that can be serviced and also upgraded based on future technology advances. And if that light source were offered to manufacturers as a drop-in-ready component, it would simplify luminaire design, fabrication and servicing, accelerating speed to market for new products. To address this need, the market is seeing significant new introductions of components called LED modules—comprehensive LED-based light sources designed for fast and easy integration by lighting manufacturers.

LED modules reduce risk for both manufacturer and user, accelerating adoption of the technology. For luminaire manufacturers, drop-in-ready modules simplify luminaire design, fabrication and servicing, accelerating speed to market for new products and providing additional freedom to innovate and differentiate their product. Users gain the benefits of a durable luminaire that can be serviced and also upgraded based on future advances in the technology. The modular approach is likely the future of LED lighting because of these distinct advantages. The alternative is to use materials that make LED luminaires easily disposable products.

Examples include Cree's LED Module LMR4, Sylvania's post top streetlight retrofit module and power supply, Molex and Bridgelux's Helieon, GE's Infusion (acquired from Journée Lighting) and Leviton's Transcend. Some modules connect the lamp module to its heat sinking and electrical supply, while others offer light source, driver electronics, optics and thermal management in a single unit. Some twist on to couple the module to the fixture, while others are designed around bases such as the GU24 bi-pin configuration. Molex and Bridgelux's Helieon LED modules, for example, which won Best of Category for Specialty Lamps and Most Innovative Product of the Year Award at 2010 LIGHTFAIR, enable beam angle, color and light output (up to 1200 lumens) to be adjusted with a single turn and click (see **Figure 5-38**).

There is currently no standard for the interface between the LED engine and the luminaire, however, so if a user buys a luminaire using a particular manufacturer's module, the luminaire may become durable, but the user will be tied to that manufacturer for upgrades and servicing. Zhaga, a European industry consortium, was created to develop standard specifications for the interfaces of LED light engines, which would enable LED engines to become interchangeable. But it is still early days for this effort.

Figure 5-37. GE's Infusion LED module installs with a clockwise twist motion that simply couples the module to its heat sinking and electrical supply. An added level of design flexibility is enabled with the module's wattage-adjust switch that gives users three different light level/wattage options. Photo courtesy of GE Lighting.

Figure 5-38. Bridgelux and Molex's Helieon modules integrate the LED source, optics and thermal management into a single assembly. The device enables beam angle, color temperature and light output—up to 1,200 lumens—to be changed or upgraded with a simple turn and click. Photo courtesy of Molex.

Interior LED Luminaires

Today, LED lighting is competitive with traditional sources in applications such as downlighting, undercabinet lighting and directional lighting. What's more, a specifier can easily locate a good solid-state luminaire, compare it with other luminaire on an apples-to-apples basis, and be reasonably confident how it will perform in the intended application (at least for the short term, as long-term *in situ* performance is still largely an unknown). The "Wild West" is being tamed, at least for those who take the time to make informed product decisions. For the uninformed, it is still easy to get burned.

In 2006, DOE began independent testing of commercially available LED products with the Commercially Available LED Product Evaluation and Reporting (CALiPER) program, confirming its suspicion that the lack of standards for testing and reporting has led to LED product performance being overstated by manufacturers. Luminaire performance,

Figure 5-39. LED luminaires are now offered for a wide range of applications, including general lighting. Photo of Stile Tryg luminaire courtesy of Stile, a brand of SPI Lighting.

for example, was being frequently represented as only the stated performance of the LEDs based on component manufacturer data, reflecting a lack of understanding by some manufacturers of their own products and/or an effort to intentionally provide misleading information to support unrealistic sales claims.

More recent DOE product testing found the fast pace of LED product introduction continuing to be plagued by inaccurate performance claims (see www.ssl.energy.gov/reports.html). Comparing lighting performance and not just watts, few LED products are competitive against the sources they are intended to replace as of the time of writing of this book. Most products are fairly new, with little time in the field. About a third of tested products in the most recent testing had accurate manufacturer claims, while another third overstated performance by 10-20 percent and the rest either did not provide performance information or grossly overstated it by as much as 100 percent.

DOE product testing, conducted in quarterly rounds, has been trending positively, however. In the past two years, efficacy doubled among tested LED products, suggestive of the market as a whole. Color quality is steadily improving. In some applications, from downlights to small-wattage replacement lamps, LED is a competitive technology. Consider downlights, an ideal application for LEDs because of the source's inherent directionality: In recent DOE testing, five 4- and 6-in. aperture, 8-39W recessed downlights were tested against 26-32W triple-tube pin-based CFL downlights. All of the LED products were found to meet or beat the light output and efficacy of 45-75W incandescent and halogen downlights, and all except one beat the efficacy levels of the CFL downlights.

Outdoor LED luminaires

One promising front in the LED revolution is outdoor lighting, which can frequently take advantage of LED lighting's unique benefits, such as energy efficiency, long service life, resistance to vibration, no mercury inside the device, aesthetic appeal, and compatibility with cold environments and lighting control strategies.

The biggest advantage that LED has over other light sources is the ability to be controlled optically towards a task. Because LEDs are directional sources, they enable design of more-efficient luminaires that place light exactly where it is needed, producing energy savings by eliminating wasteful light while potentially reducing skyglow and light trespass.

Figure 5-40. The Calculite LED downlight by Philips Lightolier, honored as "Most Innovative Product of the Year" at LIGHTFAIR 2009, provides 1,000 lumens of white light output with a choice of warm (3000K) or cool (4000K) color tones and a CRI rating of 80. The luminaire, available in 4-in. round and square models, operates at an efficacy of 50 lumens/W. Photo courtesy of Philips Lightolier.

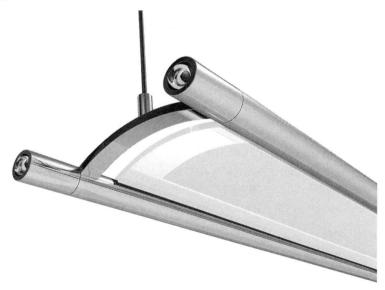

Figure 5-41. Design Excellence Award winner at the LIGHTFAIR 2009 Innovation Awards: Peerless Lighting's Kite is a direct/indirect suspended luminaire that couples LED technology with sophisticated light extractor optics; the edge-lit optic converts multiple LED point sources into uniform, glare-free illumination without a direct line of sight to the source. Each 4-ft. section produces a nominal 4,000 lumens of light output, comparable to a fluorescent luminaire with a single T5HO lamp, according to the company. Photo courtesy of Peerless Lighting.

The potential for energy savings is huge. Consider street and area lighting: According to DOE, there were some 130 million street and area lights installed in 2007, including about 48 million area lights, 46 million floodlights and 35 million street lights. If all of these luminaires were converted overnight to LED luminaires with a luminaire efficacy of 57.5 lumens/W, energy savings equal to seven 1000MW power plants, or the annual electric energy consumption of 3.7 million homes, would be realized.

Looking solely at product offerings on the market, there would appear to be no outdoor lighting category today that cannot use LEDs. Products are available for canopy, roadway, parking lot and garage, path, wall-mounted, security, landscape, bollards, floodlights, step and in-ground applications. As products continue to improve in efficacy and fall in price, the technology is moving into the outdoor market at a rapid pace.

Omnidirectional LED Replacement Lamps

While the unique characteristics of the LED source allow the lighting industry to re-imagine the luminaire, another approach is to develop

Figure 5-42. The parking lot at this St. Paul, MN Cub Foods grocery store was the first in the state to become completely illuminated by LED luminaires. Using only 0.085W/sq.ft., an average light level of 5 footcandles was achieved. Photo courtesy of BetaLED.

Figure 5-43. The Murphy USA gas station and convenience store in Plano, TX replaced HID luminaires (top) in half its facility with LED canopy luminaires. After testing the new lighting for 30 days, Murphy replaced its remaining HID luminaires with LED lighting (bottom). Photos courtesy of BetaLED.

and market integrated replacement lamps for existing luminaires. While this approach is almost never as effective as a good purpose-built luminaire, there is a huge retrofit opportunity.

Because of the Energy Independence and Security Act, more than 4 billion incandescent sockets will be up for grabs between 2012 and 2014. In residential applications, retrofit is the biggest opportunity for LEDs. Meanwhile, in commercial applications, the 2009 DOE incandescent reflector lamp regulations will declare open season on directional lighting applications, with LEDs expected to be a strong contender.

As with luminaires, LED replacement lamps are proliferating, with omnidirectional lamps, spot and flood lights, ceiling fan lamps, globes, candles and nightlights. LED replacement lamps have progressed more slowly than LED luminaires, however, which is not surprising since again, replacement lamps must compromise performance in some way to retain a conventional envelope shape and socket configuration. LED replacement lamps have made dramatic gains in just the past year, though, particularly in directional lighting, and with some promising future product introductions in omnidirectional lamps on the near horizon.

First, let's take a look at omnidirectional replacement lamps. DOE has conducted independent product testing since 2006 under the CALiPER program to help educate consumers about how LED products perform beyond the marketing hype.

Light output for decorative candelabra-type LED replacement lamps is generally comparable to low-wattage incandescent versions, with significantly higher efficacy, making this a highly competitive space for LEDs. But let's look at A-lamp replacements. CALiPER testing results up to Round 9 have been plotted on the graph in **Figure 5-44**.

For A lamp replacements, the highest-performing LED products produced light output comparable to 40W incandescents, with about 3-5 times the efficacy. CFLs, meanwhile, can produce comparable light output for even higher efficacy—approaching 60 lumens/W.

There are three things to note here. First is that products show a wide range of performance. Second is what's not shown here—that the tested replacement lamps also showed wide variations in other performance attributes such as color quality, form factor, light distribution pattern and power factor that would disqualify some of the products from many if not all applications. But let's be fair. The third thing to note is that the information shown here is dated, showing results from 2006 to 2009, with each report having a lag of several months. Product performance is

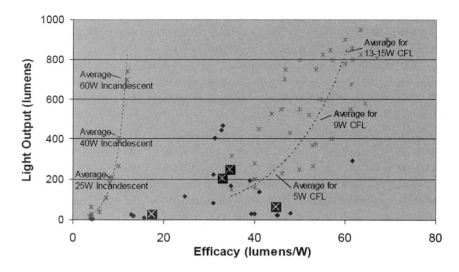

⊠	SSL A-lamps Round 9, 6-7W
⊠	SSL Candelabras Round 9, 1-2W
♦	SSL CALiPER Earlier Tests, 1-14W
✕	Incandescent Benchmarks, 4-60W
✕	CFL Lamp Benchmarks, 4-18W

Figure 5-44. CALiPER testing results for omnidirectional LED replacement lamps.

advancing steadily in the meantime.

To promote product development to replace the most common household light bulb, the Energy Independence and Security Act of 2007 created the "Bright Tomorrow Lighting Prizes," or the L Prize, for LED replacements for today's 60W incandescent general-service lamps and PAR38 lamps, with a $10 million cash prize for the first lamp that meets stringent requirements. The 60W replacement must have an efficacy of higher than 90 lumens/W while producing more than 900 lumens, drawing less than 10W of power, and providing a 90 CRI and rated life of 25,000 hours. Philips submitted the first entry, which was being tested at the time of writing. The product made Time Magazine's Best Inventions of 2009 list. But even if Philips wins, the contest is not over, as three winners are eligible for Federal purchasing opportunities and promotions and incentives offered by the L Prize partners.

The Big Three lamp manufacturers—GE, Philips, Sylvania—began offering omnidirectional LED replacement lamps in 2009-2010. For example, Philips announced the Endura A55 AmbientLED, a 7W accent light designed to replace the 25W incandescent lamp in sconces, fan lamps and hospitality applications. At the Light+Building show in Frankfurt in 2010, Philips announced that by the end of the year it would be launching a Master LED dimmable 12W lamp that at 806 lumens would be able to replace 60W incandescent lamps. GE's Energy Smart lamp is a 9W LED replacement fitted with Cree X-lamp XP-G LEDs, designed to replace 40W lamps (see **Figure 5-45**). Sylvania announced that it would have a replacement for 60W A-lamps in September 2010.

Meanwhile, smaller companies are also innovating. Ledzworld Technology, for example, won a Best of Category award at LIGHTFAIR 2010 for its Professional LED Bulb CTA (see **Figure 5-46**). The product uses Color Temperature Adjusted (CTA) dimming, which enables the lamp to gradually change from a warm color tone to an even warmer flame color, similar to the way an incandescent performs while it is being dimmed.

Figure 5-45. GE's Energy Smart 9W LED replacement lamp is designed to replace 40W incandescent lamps. Photo courtesy of GE.

Figure 5-46. Ledzworld Technology's Professional LED Bulb CTA, which shifts from a warm color to an even warmer flame color during dimming, similar to an incandescent lamp when it is dimmed. Photo courtesy of Ledzworld.

Directional LED Replacement Lamps

LED technology holds significant promise for directional lighting applications. LEDs are an inherently directional light source, resulting in efficiency gains in directional lamps, and emit no radiated heat and UV radiation, making them ideal for museum, retail and supermarket applications. Viable offerings include MR16, PAR16, PAR20, PAR30, PAR30 Long Neck, PAR38 and other large sizes such as BR40 for replacement of halogens up to 45W for warm-white (2700K) products and 60W for cool-white (4200K) products. Products are available with flood and wide flood distributions, suitable for general lighting, and spot distributions suitable for accent lighting.

DOE CALiPER testing of light intensity is shown in **Figure 5-47**. Center beam candlepower (CBCP) is the intensity in candelas emitted at the center of a directional lamp beam (0°, or nadir). Beam angle is the angle at which the beam intensity is 50 percent of the CBCP and defines whether the lamp has a narrow, medium or wide distribution. As a rule of thumb, a lamp with a large beam angle will have a lower CBCP than a lamp with narrow distribution, as shown on these curves.

As with omnidirectional lamps, performance can vary widely from one manufacturer product to the next. The test data shows that performance is improving, as evidenced in the latest Round 9 CALiPER testing.

Some examples of recent product introductions include GE's Energy Smart LED PAR20, designed to replace 30W R20 incandescent lamps. This product has replaced some 12,000 incandescent and halogen lamps in more than 150 Red Robin restaurant locations in the United States (see **Figure 5-48**). Nexxus Lighting's Array Quantum LED R30 replaces the traditional phosphor over LED approach with a quantum dot material, developed in conjunction with QD Vision, which allows more precise color tuning; the result is a warm white directional LED R30 with a high CRI of 91 and initial efficacy of 60 lumens/W based on LM79 testing (see **Figure 5-49**). Lighting Science Group's SoL dimmable R30 can be dimmed to less than 1 percent light output, according to the company. And Sylvania introduced at LIGHTFAIR 2010 an LED PAR38 with a broader choice of narrow beam angle options than conventional halogen PAR lamps.

T8 LED Replacement Lamps

There is currently a push to replace linear fluorescent T8 and T12 lamps with linear LED replacement lamps, but are LEDs really ready to take on this workhorse in general lighting? The evidence suggests no, not yet anyway, as of the time of writing—except perhaps in some specialty applications such as refrigerated display cases.

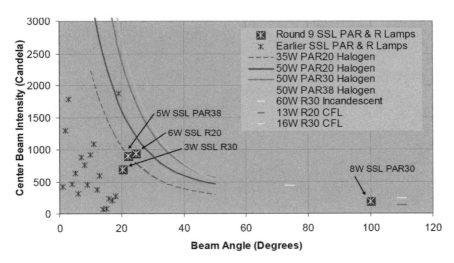

Figure 5-47. **Minimum center beam intensity for PAR and R lamps in DOE CALiPER testing.**

Figure 5-48. PAR20 LED lamps replaced some 12,000 incandescent and halogen lamps in more than 150 Red Robin restaurant locations in the United States. Photo courtesy of GE.

Figure 5-49. Nexxus Lighting's LED R30 uses quantum dot technology to achieve precise color and a warm white directional LED lamp with 91 CRI and initial efficacy of 60 lumens/W. Photo courtesy of Nexxus Lighting.

Linear LED replacement lamps are designed to directly replace 4-ft. T8 and T12 lamps in existing luminaires. A random sampling of products shows wattages ranging from 15W to 25W, availability in 2- and 4-ft. sizes, selection of color temperatures from warm white to daylight, and ability to operate with or without the existing ballast. A service life of 50,000 hours is usually claimed. Here are four 4-ft. products introduced at 2010 LIGHTFAIR:

- 17W T8 producing 1,200 lumens;
- 18W T8 producing 1,540 lumens (warm white), 1,695 lumens (neutral white) or 1,8-50 lumens (cool white);
- 19W T8 producing 1,650 lumens of 4500K or 6000K light; and
- 25W T8 producing 1,600 lumens of 2700K or 4100K light.

According to manufacturers, advantages include energy savings, long service life, suitability for cold-temperature applications, directional light, no mercury or lead, and resistance to shock/vibration. Today's white-light LEDs, for example, can produce up to 50,000 hours of service life, with few spot replacements. LEDs also produce light directionally instead of in all directions like a fluorescent tube, which can increase the efficiency of luminaires into which they are installed.

LED manufacturers may also claim that even though their products emit fewer lumens than the T8 lamps they are intended to replace, their LED lamps are actually "brighter." This is because LED lamp manufacturers exploit the directionality of LEDs to emit light from the T8 tube in a downward direction instead of all directions, as a conventional T8 lamp would, thereby improving luminaire efficiency as less light is trapped inside the luminaire and wasted.

But do LEDs provide sufficient value in this application? At $45-$150/tube, they aren't cheap, particularly when the competition—the T8 lamp—can be acquired for less than $5. Suppose we have a two-lamp luminaire with fluorescent T8 lamps powered by an electronic ballast, drawing 58W. The luminaire operates 3,120 hours/year with a utility energy cost of $0.10/kWh. After replacing the two lamps with 17W LED lamps, we save 24W. The savings for this luminaire are $7.50 per year. If the cost for each LED lamp is even as low as $45, the payback based on product cost alone is 12 years. At $100, it's 13 years. At $150, it's 20 years.

To sweeten the deal, we could include estimated lamp replacement and labor savings because of the maintenance benefit, but remember there are extended-life fluorescent lamps offering rated service life of 40,000 to 46,000 hours (at 3-12 hours/start on a programmed-start ballast), approaching the best that white-light LEDs can currently do. Additionally, many white-light LED devices actually produce less than 50,000 hours of useful life due to overdriving the LEDs to produce more light output. That 50,000-hour is not a magic number, but rather represents typical performance among the best-designed LED products and applications. All things considered, the maintenance argument is less decisive than would first appear.

The main problem, however, is low light output and resulting low efficacy. CALiPER-tested LED T8 replacements were found to produce only a third of the light output of the lamps they are intended to replace (see **Table 5-8**). Median efficacy for LED T8 products tested to date is 44 lumens/W (the highest was 70 lumens/W). While it is true that LEDs are directional and therefore emit light more efficiently from the luminaire, the improvement was found to be not enough for equivalent performance. The result is that more lamps would be needed to maintain the design light level. What's more, the directionality of the LEDs was found to result in narrower distribution from the luminaires, which affected uniformity and may produce dark spots in what was before a uniformly lighted space. In addition, these lamps bypass the ballast or use the existing ballast in retrofits. If the ballast is bypassed, more labor is required. If not, efficacy is reduced. Finally, DOE also found that color temperature was frequently misstated, with two tested LED replacement products found to have color temperatures well over 6000K—one 7739K and the other 12583K—much "colder" than recommended.

DOE concluded: "In general, solid-state linear replacement products are not competitive in troffers as a replacement for linear fluorescent lamps at this time."

As a result, LED T8 replacement lamps were not included in the ENERGY STAR criteria for LED replacement lamps that became effective August 2010. (ENERGY STAR is described in detail later in this chapter.) DOE has, however, published recommended specifications for 4-ft. LED linear replacement lamps to provide a performance level that is reasonable to demand (see **Table 5-9**). Be sure to request LM79 photometric test

Table 5-8. CALiPER test results for 4-ft. LED T8 replacements with fluorescent benchmarks. Source: DOE.

Performance Characteristics	LED T8 Replacements		Fluorescent Benchmarks	
	Range (12 lamps tested)	Average	Mfr. Data (75 lamps)	CALiPER (2 fixtures tested)
Initial Lamp Light Output (lm)	345 – 1,579	1,111	2,778	3,091
Initial 2-Lamp System Efficacy (lm/W)	19 – 76	50	87	---
Initial 2-Lamp Fixture Light Output (lm)	597 – 2,038	1,563	3,577	4,064
Initial CRI	63 – 76	71	75 – 80	82
Fixture Efficiency (%)	74 – 86	83	74	66
Initial Luminaire Efficacy (lm/W)	17 – 57	41	64	57

Table 5-9. Suggested specifications for 4-ft. LED linear replacement lamps. Source: DOE.

Performance Attribute	Specification	Notes
Initial minimum lamp light output *LED useful life is based on 70% lumen maintenance. Fluorescent T8 lamps have much higher lumen maintenance, about 94%. The initial lumen specification for LED T8 aims to ensure equivalent output to fluorescent T8.* *Measured fixture efficiency with LED T8s averaged 17% higher than with fluorescent, due to LED directionality.*	2,700 lumens	Based on equivalency to fluorescent T8 (average of 75 T8 lamps) with the following: – Average initial fluorescent T8 lumens = 2,778 – Normal ballast factor = 0.87 – Depreciation of fluorescent lamp lumens at time of relamping (70% of rated life) = 0.94 – Fixture efficiency factor, fluorescent compared to LED = 0.17 – Fluorescent depreciated lumens = 1,886 – Divide by 0.7 to get LED initial lumens = 2,694, round to 2700
Minimum lamp life, L70 *Unlike for conventional light sources, there is no standard life rating method for LED T8 lamps. The products are too new to have long-term operating data available.*	35,000 hours Ask for 6000 hours of integral lamp operating data (not just LED data). Lumen maintenance at 6000 hours should be at least 94.1% of initial value.	Fluorescent T8 rated life averages 24,000 hours on 3-hour starts and 30,000 hours on 12-hour starts, both on instant start ballasts. Longer life of 40,000 hours or more is possible with programmed start ballasts.
Luminous intensity distribution	Varies, but users are strongly encouraged to evaluate intensity distributions in comparison to those from fluorescent systems. If possible ask for an intensity distribution plot for the LED T8 lamps in the intended fixture type (e.g., parabolic, lensed, etc).	
Correlated color temperature (CCT) (Kelvin)	Nominal CCTs and tolerances as defined in ANSI_NEMA_ ANSLG C78.377-2008, "Specifications for the Chromaticity of Solid State Lighting Products." | Nominal CCT	Tolerance
---	---
2700 K	2725 ± 145
3000 K	3045 ± 175
3500 K	3465 ± 245
4000 K	3985 ± 275
4500 K	4503 ± 243
5000 K	5028 ± 283
5700 K	5665 ± 355
6500 K	6530 ± 510		Chromaticity tolerances defined in ANSI C78.377 correspond to approximately 7-step Macadam ellipses for fluorescent chromaticity. In addition to the six nominal CCTs defined for fluorescent sources, the standard defines 4500K and 5700K for SSL products.
Minimum CRI	80	Equivalent to 800 series fluorescent lamps.
Warranty	3 years	
Electrical Safety	ANSI/UL 8750	

reports from a NVLAP-accredited or CALiPER-qualified testing lab (see www.ssl.energy.gov for more information) and lumen maintenance testing on the full lamp (at least 6,000 hours is recommended by DOE).

There are several important lessons here: First, just because a lighting product is LED does not automatically mean it's efficient. Second, the best defense against misleading sales claims is knowledge. Third, trust sales claims for any lighting product only if they are backed by independent, reputable test results. And fourth, testing a retrofit lamp yourself in the intended application prior to purchasing is always a good idea.

In the end, LEDs really are just another light source and, as with all light sources, if they produce a benefit—in terms of either lower cost of operation, enhanced operational serviceability or lighting quality—that is equal to or greater than their cost, then they are a good choice. If not, then you are taking a risk.

In the Spotlight: OLEDs

At this point, it is easy to imagine an LED future. But another solid-state light source is emerging that may revolutionize lighting even further. It's called organic light emitting diodes, or OLEDs—thin, organic materials sandwiched between two electrodes, which illuminate when an electrical charge is applied. In addition to widespread design capabilities, OLEDs have the potential to deliver dramatically improved levels of efficiency and environmental performance, while achieving the high quality of illumination found in traditional LED systems.

At the 2009 Euroluce International Lighting Fair in Milan, Royal Philips Electronics unveiled its first organic LED (OLED) interactive lighting concepts designed for both consumer and professional use. The concepts offer a combination of intuitive interactivity, ultra flat shape, soft light-effect and new design possibilities. Philips showcased four concepts—standing, wall, desktop and ceiling luminaires—each incorporating flat OLED light panels supplemented with Luxeon Power LEDs for the functional lighting component (see **Figure 5-50**).

In the fall of 2009, OSRAM Opto Semiconductors announced that they had developed ultrathin transparent OLED panels. This enables a light source that is in plain sight and yet virtually invisible until it is ON.

And GE recently announced a series of OLED luminaire concepts developed using thin-film technology from Konica Minolta (see **Figure 5-51**). GE is focusing on a low-cost roll-to-roll production process similar to newspaper printing, which would enable economical manufacturing of OLED film.

Why is this technology important? Imagine a future where light sources

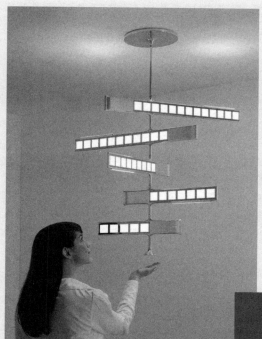

Figure 5-50. OLED concept luminaire. Photo courtesy of Philips.

become pixels of information ... windows are transparent during the day, allowing daylight to enter the space, and then become brightly luminous at night ... rooms are lighted using luminous wallpaper and room dividers ... and where people wear luminous safety clothing, where lighting is completely integrated into furniture and window coverings and building materials, and where whimsical luminaires redefine our concept of a light source.

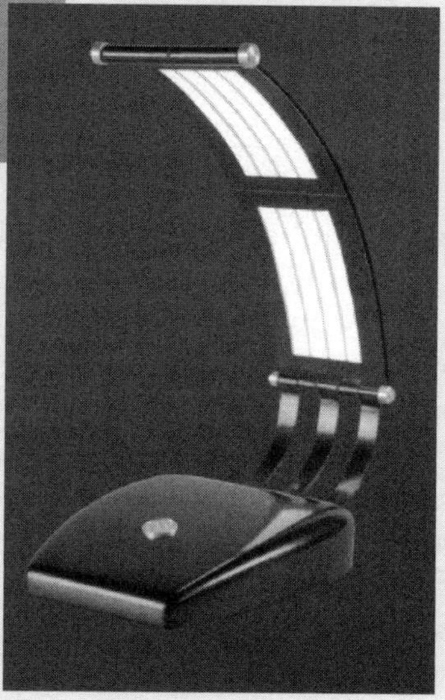

Figure 5-51. OLED concept luminaire. Photo courtesy of GE.

Figure 5-52. Concept view of OLED lighting integrated into stairs. Photo courtesy of GE.

Understanding LED Service Life

As LEDs find an increasing number of niche commercial general lighting applications where they can be competitive with traditional light sources, understanding and accurately representing service life for these devices becomes more important.

In short, LED lighting products do not fail similarly to traditional light sources such as incandescent and fluorescent. Fluorescent lamp life, for example, is expressed as the mean time between failures—the point at which 50 percent of a large population of lamps is expected to fail. If LED lamp life were expressed using this metric, the service life of some LED products might be 100,000 hours or even longer. However, by 100,000 hours, it is likely that the useful life of the product would long be over.

Instead, the useful life of LED products is based on lumen depreciation. All light sources experience a gradual loss of light output over their lives. High-performance T8 lamps provide good lumen maintenance. LEDs, it can be argued, do not. The LED product may continue producing light for 100,000 hours or more, but by the time it fails it will have experienced significant losses of light output to an extent that it was no longer economical to continue operating long before that point.

The Alliance for Solid-State Illumination Systems and Technologies (ASSIST), formed by the Lighting Research Center in 2002, recommends defining the life of LED products based on lumen depreciation, a convention that has been accepted by the industry as a de facto standard. For example, ASSIST recommends 70 percent lumen maintenance (30 percent loss of light output) as useful life for general lighting applications (L_{70}), as research suggests a majority of occupants will accept a 30 percent reduction of initial light output while not noticing the gradual light loss up until this point. For today's best white LED sources, this is about 50,000 hours (see **Figure 5-53**). If light output is not critical, such as in some decorative applications, 50 percent lumen maintenance is recommended (L_{50}). However, cold or hot junction temperatures in the LEDs can respectively extend or shorten useful life, with temperature affected by a range of application factors.

The question becomes: How will the owner know an LED product has "failed" when it is continuing to produce light? After all, even a 50,000-hour service life is very long compared to traditional sources—13 years in an average office building. With the added maintenance benefit of there being no spot relamping between failures, LED products are virtually "install and forget" devices. In many applications, maintaining light levels is vital for productivity and safety, particularly when minimum light levels are established by code; down the road, this could become a problem. After more than 10 years in operation, will anyone think of checking light levels and replace products that are apparently still functioning? What happens if the luminaires are allowed to operate well beyond the design luminance levels in the case of code-required minimums?

One idea is to force an end-of-life action in LED products. For example, LED luminaires could be developed with a timeclock that signals automatic shutoff, alarm or enunciator light to indicate that designated number of hours has transpired. This is a simple solution that provides actual shutoff or a signal to users that they need to replace their luminaires, but does not address the fact that LED lumen depreciation may vary widely based on application characteristics. Another solution might be an active feedback device that monitors light output continuously and either shuts off the luminaire or sends an alarm signal or turns on an enunciator light to indicate the luminaire's useful life

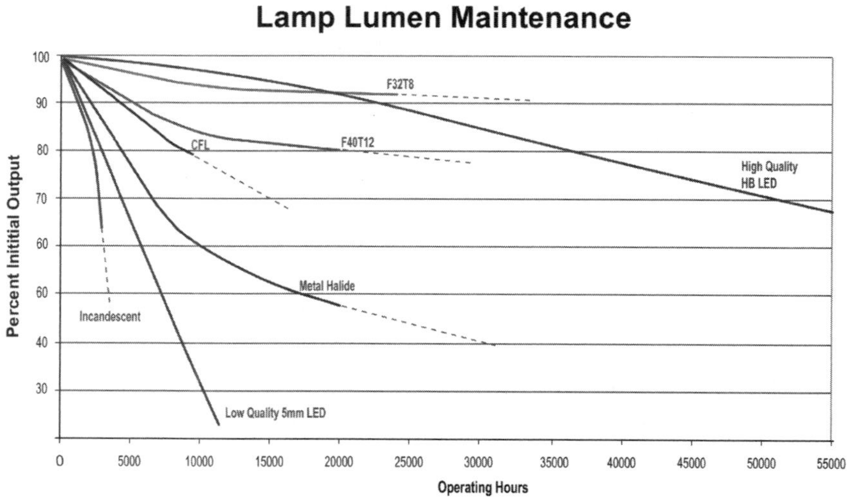

Lamp Lumen Maintenance

Figure 5-53. Comparison of lamp lumen depreciation rates for popular lamp types including high-quality high-brightness LEDs. Data compiled from OS-RAM SYLVANIA, Dialight and Nichia. Graphic courtesy of Kevin Willmorth.

is over and the product needs to be replaced. However, these features present a premium, and are unlikely to be included in LED products unless end-users demand it.

In an LED future, another solution might be for a lighting management company or some other "lighting steward" to monitor lighting performance by periodically testing light levels. When light levels drop below the design level, the LED lighting would be replaced.

Lighting Controls and LED Lighting

Combining solid-state building illumination (such as LED) with intelligent, automatic controls in the future can accelerate energy savings and extend LED service life. LEDs, in fact, are ideally suited for lighting control in a number of ways.

Switching, for example, is friendly to LED lighting, which can be controlled using both manual switches and automatic switches such as occupancy sensors and photocontrols. LEDs turn ON and achieve full brightness instantly, and frequent switching does not affect lamp life, unlike fluorescent and HID lamps. Be sure to avoid low-end occupancy sensors and photocontrols that may not work well with LED devices.

Figure 5-54. Target Corporation is installing LED lighting over 55,000 doors in refrigerated display cases in 500 stores across the United States, reducing energy consumption by an estimated 60 percent. LEDs provide a number of significant advantages over fluorescent in cold environments, including reliable starting and instant-ON operation, energy efficiency and long service life that thrives on lower ambient temperatures. Photo courtesy of GE.

Dimming is also considered very friendly to LED lighting. Like fluorescent dimming, a dimming range is available from 1-20 percent to 100 percent, depending on desired capabilities and acceptable cost. The highest-performing products dim smoothly from 1-100 percent, while the lowest dim from 20-100 percent while exhibiting an inconsistent step-dimming effect.

Unlike conventional fluorescent dimming such as 0-10VDC, which experiences a degradation of efficacy accelerating towards the low end of the dimming range, LEDs produce light output that is proportional to electrical input. An LED operating at 80 percent of its initial rated power will produce roughly 80 percent of its initial rated light output.

Towards the low end of the dimming range, however, a remarkable reverse effect occurs as happens with fluorescent—efficacy actually increases. This is because LEDs are very sensitive to internal temperature, and as they are dimmed, temperature drops, resulting in an increase in light output. This is good for energy management, but for manual dimming, it means light output and dimmer setting may drift out of proportion at the low end of the dimming range. Some higher-end products compensate for this.

A reduction in thermal stresses on LEDs due to dimming can have other positive effects. As noted above, reductions in LED internal temperatures will increase light output. Since service life with white LED products is based on light output—with useful life defined for general lighting application as the point at which the light source is emitting 70 percent of its original light output—dimming can increase LED service life. Additionally, high operating temperatures can cause a color shift towards blue among most white LEDs as their phosphors fail. By reducing temperatures, this color shift can be delayed. (Note that if LED luminaires are layered in a space in two or more separately controlled control zones using today's popular technology, the result may be color inconsistency developing between luminaires in each control zone.)

Today, however, controlling white LED products in architectural lighting applications can be a challenge. Some products have a driver and power supply that enable dimming, others do not. Many are proprietary and incompatible with products from other manufacturers. Some do not work with basic electronic thyrister- or triac-based sensors and relays. Some do not work well when connected to standard dimmers or existing 12V electronic transformers.

When dimming—or the ability to integrate the LEDs into a cen-

Figure 5-55. LED controller for select LED drivers/sources in high-end applications. Photo courtesy of HUNT Dimming.

tralized control system—are desired, special LED drivers and power supplies are typically needed. The driver regulates current flowing through the LED devices just the way a ballast regulates current flowing through a lamp. Sometimes, the driver and power supply are integrated into a single product. These devices can be mounted remotely from the LEDs. The driver is the point of connection between a LED lighting system and its controller. Lighting and controllers can be connected via 0-10VDC analog (typical for white LEDs), digital (developing for white LEDs) or DMX512 (typical for color LEDs) communication methods. When selecting 0-10VDC controls, note that not all controls definitively shut off the power in the off state, and require a separate switch or relay to do so.

For retrofit, there are line-voltage AC LED products that do not use driver circuits or power supplies, and work directly on 120VAC power and with existing dimmer controls that are rated as compatible by the manufacturer (but beware of flicker). Additionally, 12VAC products are available that operate with dimmable 12VAC transformers (but ensure the transformer is compatible with other controls and sensors).

As LEDs are still a young technology and standards are still being developed by NEMA that will provide metrics for evaluating LED dimming compatibility and performance, selecting LED replacement lamps and matching them with existing dimmers should be approached with caution. *LED lamps and controls should be verified as compatible (assume they are not until proven otherwise).* Any chosen product should meet safety performance requirements as well as power quality and industry standard specifications. The driver should have a rated life comparable to the LED array (20,000 to 50,000 hours). If layering dimmable and non-dimmable LED lighting in the same space, these different systems may age differently, resulting in diverging light output levels and degree of color shift. Because light output increases at the low end of the dimming range, the light output and dimmer setting may drift out of proportion at this low end. And because incandescent line-voltage dimmers are designed specifically for incandescent lamps, their requirements—such as constant leakage current path, minimum load, resistive impedance, which are met by most incandescent lamps—must be met by the given LED product, or else the product may flash, flicker, not turn ON or operationally fail.

LEDs: TAMING THE WILD, WILD WEST

CFLs in America: Lessons Learned on the Way to Market, published by DOE, concluded that technical and quality problems with early compact fluorescents snowballed into major obstacles to acceptance by consumers and retailers.

Without standards and testing, products may be misrepresented by unclear or exaggerated claims about performance. Without uniform terminology to describe performance, distributors and consumers can become easily confused. And if distributors and consumers get a negative initial impression of a technology, it may last for decades.

These lessons were not lost on DOE, which is now working hard to support adoption of LED lighting, an emerging technology that is experiencing some of the same pains.

The biggest problem affecting all LED products continues to be inaccurate performance claims. In recent DOE CALiPER testing, about a third of the tested products were backed by accurate claims while another third overstated performance by 10-20 percent and the rest either did not provide performance information or grossly overstated it by as much as 100 percent. It is this atmosphere—in which LED products vary widely in performance, few products are competitive against the sources they are intended to replace, and most products are fairly new, with little time in the field—that has prompted more than one comparison to the LED market as the WILD, WILD WEST.

To support adoption of solid-state lighting, DOE has engaged the lighting industry in several meaningful ways, from fast tracking solid-state lighting testing standards to launching ENERGY STAR labeling criteria for LED luminaires and replacement lamps to reporting independent testing data on LED products on the market.

The idea is to minimize risk and build confidence among specifiers and users while rewarding quality by recognizing it.

The infrastructure is now being put into place that enables lighting designers to tame the Wild West—at least for those who take the time to get educated. Those who do not, again, are taking their chances.

Standard Performance Testing

To advance market adoption in general lighting, lighting customers at all levels need to readily understand what a specific LED product does and have confidence that its represented performance will be delivered as promised. This includes confidence that product claims are based on accepted testing methods that enables apples-to-apples comparison. Further, there must be confidence that performance will be consistent and dependable.

This is why development of robust standards is critical for the solid-state lighting community: Standards eliminate inconsistencies in representation of product attributes such as color, intensity and service life. They also help buyers and specifiers assess and compare product performance based on common, industry-accepted and meaningful metrics.

Based on more than a century of experience, testing standards for conventional light sources are readily available. Unfortunately, these do not always work well when applied to solid-state lighting products. Due to the unique characteristics of LEDs, conventional testing processes fail to capture important performance differences. For example, operating temperature ranges that have little impact on incandescent lamp life can have a major impact on the life of an LED product. Additionally, the rapid rate of change in LED technology makes it impossible to base test protocols on standardized light sources, as there are none at this time.

The basic metrics for assessing performance of solid-state lighting products remains essentially the same as any other source. Specifiers' need to know a product's light output, input watts, efficacy, color quality and expected service life is no different than for any other source. However, until 2008, no industry-accepted standards existed for testing and representing these basic product attributes for LED lighting products. The use of relative photometry—where luminaire performance is measured, the lamping and ballasting are removed and tested separately to establish an efficiency multiplier, and the photometry report is created using a standardized lamp lumen rating—simply does not work in LED products whose sources are integral to the luminaire and are non-standard in configuration. Further, the effect of thermal management—most LEDs cannot be tested without effective heat sinks attached—along with integral electronic driver components makes separation of "lamp" and luminaire impossible. This means that relative photometry simply cannot be applied. Applying conventional service life calculations based on 50 percent of lamp failures (to determine average life) is also unusable for sources that rarely fail outright but exhibit continual lumen depreciation that renders them unusable before they expire.

The first standard established for LED products addressed the service life rating issue. The Alliance for Solid State Illumination Systems and Technologies (ASSIST) established two metrics for representing useful life based on lumen depreciation. For commercial products, 70 percent of initial lumen output, and for residential or non-critical applications, 50 percent of initial lumens, provides a foundation for evaluation. However, without test standards to establish these numbers, they are virtually meaningless.

To address this, IES LM-80-2008 prescribes the test methodology for lumen depreciation for LED-based packages, arrays and modules. This test standard establishes test conditions—including three different operating temperatures to simulate likely field conditions—and equipment requirements. The lumen output of the LED device is measured every 1,000 hours to at least 6,000 hours (10,000 hours preferred). Based on this information, lumen depreciation can be extrapolated to the L_{70} or L_{50} value. IES TM-21, a procedure in development by IES, will soon outline specific procedures for deriving the LM-80 test data into service life rating, establishing consistency between all manufacturer data.

Recognizing the need for rapid deployment of useful standards, the IES fast-tracked the development of LM-79-2008, Electrical and Photometric Measurements of Solid-State Lighting Products. DOE supported this accelerated process to address its concern that inaccurate information in the marketplace could result in the same dissatisfaction among consumers (and stunted market acceptance) experienced with compact fluorescent lamps in the residential market. The establishment of standards is regarded as essential to avoid inhibited adoption of solid-state lighting in general illumination applications.

Standard LM-79-2008 includes the environmental conditions for testing, how to operate and stabilize the LED sources during testing, and the methods of measurement and types of instruments to be used. It is based on a method called absolute photometry, which simply means that the reported information is directly representative of the specific product tested and is not derived relative to any rated lamp standard. Essentially, the product is measured using an integrating sphere to measure gross light output and color characteristics, and a goniophotometer, a photometer that measures light intensity reflected from a surface at different angles (traditionally used for conventional luminaires to measure luminous intensity distribution values, but using relative photometry). The combined testing objectively captures 11 aspects of performance including total lumen output, luminaire efficacy, chromaticity(from which color rendering index, or CRI, rating can be derived), and correlated color temperature. Product performance data provided by manufacturers that have tested their products using LM-79 creates needed reliability of information for decision making and comparison.

Without this, there is no way of knowing whether performance information provided represents the luminaire itself or just the LED source manufacturers data generated in isolation of any actual product application.

ENERGY STAR for Luminaires

To strengthen user confidence in this emerging technology, DOE established ENERGY STAR criteria for LED luminaires that took effect September 30 of 2008. ENERGY STAR is a voluntary labeling program designed to help consumers identify energy-efficient, cost-effective products on the market. In the case of LED luminaires, it can help consumers easily identify energy-efficient, high-quality LED products that perform as well or better than similar product using traditional light sources.

LED luminaires that meet the strict criteria can be labeled with the ENERGY STAR mark, which indicates the product not only meets a certain threshold for efficiency, but also performance. ENERGY STAR criteria now cover undercabinet kitchen and shelf-mounted task, portable task, recessed downlight and outdoor wall-mounted porch, step and pathway luminaires, surface and pendant downlights, ceiling luminaires with diffusers, cove lighting, surface luminaires with directional heads, outdoor pole/arm decorative luminaires, wall washers and bollards. These categories will continue to expand as LED technology improves to a point where it is competitive. Ultimately, all general lighting categories will be covered.

As of the time of writing, ENERGY STAR commercial market products have been introduced by Cooper, Cree, Digital Lighting, EEMA, Elite, Intematix, Juno, Lithonia, Philips, Prescolite and Renaissance Lighting.

Figure 5-56. Madison Area Technical College installed 400 ENERGY STAR-qualified LED downlights across their campus, including the cafeteria shown here. Photo courtesy of Cree, Inc.

Figure 5-57. British Columbia's Delta school district replaced seventy-two 200W incandescent recessed luminaires with an equal number of LED luminaires operating less than 15W on a one-for-one basis, generating 93 percent energy savings and a payback of 2.4 years after a utility rebate. Photo courtesy of Cooper Lighting.

As of the time of this writing, more than 160 products have become EN-ERGY STAR-qualified, mostly recessed downlights but with some surface and pendant downlights and undercabinet shelf luminaires as well. For a complete, current list, visit www.energystar.gov and type "SSL" into the search box.

ENERGY STAR for Replacement Lamps

On August 31, 2010, the final ENERGY STAR criteria for integral LED replacement lamps became effective. In the case of LED replacement lamps, the criteria require the product to be at least as efficient as comparable compact fluorescent lamps while providing light output, color and distribution equivalent to incandescent or halogen lamps.

The criteria cover lamps in non-standard forms (which may not claim equivalency with any standard lamps) as well as in forms intended to replace existing standard electric lamps (which may claim equivalency). Criteria covering the latter include omnidirectional (light emitted in all directions), directional (80 percent of light output falling within a 120-degree cone) and decorative lamps (globes, candles, etc.). Let's look at omnidirectional and directional lamps.

Omnidirectional lamps, including A, BT, P, PS, S and T lamps, must operate at a minimum efficacy of 50 lumens/W (<10W LED lamps) or 55 lumens/W (≥10W). The lamp must produce light output based on the wattage intended for replacement, expressed on a scale starting at 200 lumens (being able to claim replacement for a 25W incandescent) and running up to 2,600 lumens (150W replacement). Not surprisingly, the lamp is expected to produce a uniform 360-degree distribution.

Directional lamps, including BR, ER, K, MR16, PAR16, PAR20, PAR30, PAR30LN and PAR38 lamps, must operate at a minimum efficacy of 40 lumens/W for ≤2.5-inch-diameter and 45 lumens/W for >2.5 inch-diameter lamps. The lamp must produce a minimum light output of 10 times the wattage of the lamp intended for replacement. Determining the equivalent conventional directional lamp is based on entering wattage and beam angle into a center bream candlepower calculator (note for several reasons it is not a perfect equivalent; performance may still vary).

For both omnidirectional and directional lamps, the lamp's color tone must be 2700K, 3000K, 3500K or 4000K; exhibit a certain level of color maintenance; and have a minimum CRI rating of 80, with R9>0. If the lamp is >5W, it must have a power factor ≥0.7. Finally, it must have a rated service life at least 25,000 hours and have basic information published in a standard Lighting Facts label on the product packaging.

What is R9 and why is it important? CRI is based on an average of the R1 through R8 color palette. R9 through R14 are saturated colors; R9 is the saturated red. Halogen sets the bar with an R9 of 100. A high R9 is important for rendering red well in display lighting in food, flower, clothing and similar applications. Many new ceramic metal halide products commonly used in retail display lighting, for example, claim high R9 values.

What about dimming? As stated earlier in this chapter, dimming reduces LED internal temperatures, which can increase efficacy and extend product life. However, not all LED replacement lamps are dimmable, and not all dimmable products are compatible with all existing dimmers. DOE is currently working with NEMA on a dimming standard. Until then, ENERGY STAR only requires that the manufacturer clearly state whether the product is dimmable on the product packaging and then publish information on a web page about dimmer compatibility.

What about LED linear T8 replacements? As stated earlier in this chapter, as of June 2010, DOE has bluntly stated that these products are not ready for prime time based on product testing in which LED lamps were found to produce only a third of the light output of the lamps they are intended to replace.

Lighting Facts Labeling

In late 2008, DOE and the Next Generation Lighting Industry Alliance launched the Quality Advocates Initiative. Participating manufacturers voluntarily pledge to follow certain guidelines and labeling to report performance claims for their products. To facilitate demand for quality products, the SSL Quality Advocates Initiative also registers lighting professionals and distributors/retailers.

Each product is tested under the regime defined by the IESNA LM-79-2008 standard described earlier in this chapter. Once the test data is verified and the product completes its registration, the product can carry the Lighting Facts label (see **Figure 5-58**). This label provides—in a quick, simple, consistent format similar to nutrition labeling—summary performance data covering light output (lumens), power (W), efficacy (lumens/W), correlated color temperature (K) and CRI rating.

This simple convention is important because it promotes and rewards more accurate and consistent reporting of product performance. It empowers those who recommend, sell and install lighting to easily com-

---

In the Spotlight: Reducing the Risks that Come with LED Lighting

LED is becoming competitive against conventional technologies in a growing list of applications, but despite all the progress, it is still emerging. As product performance improves, high initial cost and undercutting by cheap, poor-quality products tainting the marketplace will remain the biggest inhibitors to adoption of LED lighting.

In "Solid State of Lighting," an article the author wrote for Electrical Contractor Magazine's March 2010 issue, Kevin Willmorth, principal of consulting firm Lumenique, LLC, provides the below advice for practitioners interested in reducing the risks that come with LED lighting:

- Look for the highest-quality product the project can afford, and avoid cheap products completely. This is a get-what-you-pay for time for LED lighting; be prepared to spend more to get more.
- Demand details on color quality and life ratings. If the manufacturer does not know, do not buy from them.
- Test products and ask for demos. This is the first step in avoiding the hyped-up promises that produce poor-performing products.
- Avoid retrofitting incandescent sockets if possible. This looks like an easy way to get started, but is also an easy way to end up with poor performance or ugly results.
- Believe no claim that products can be controlled by any dimmer. If dimmers will be used, test the combination of the LED lighting and given dimmer control.
- Pay attention to thermal issues such as insulation and air circulation. Unless the manufacturer specifically states the product is suitable for an intended application, assume it is not.
- Keep LEDs in context. There is no magic here. If they work for a given application, use them, otherwise don't.
- Stay informed and engaged. Try a few samples, get to know the technology.

---

pare products. And quite simply, it reminds specifiers and end-users to consider a larger story for the product than just its wattage.

It does not tell the whole story of the product, however. For example, the Lighting Facts label currently does not provide a service life rating (hours). It does not state whether the product can be dimmed and if so what types of dimmers it is compatible with. It does not say whether the product is suitable for indoor or outdoor use, or both. And it does not state whether the product is suitable for installation in enclosed spaces

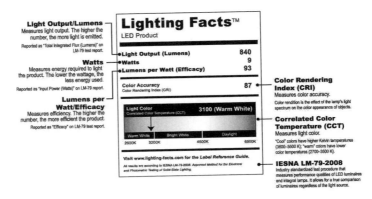

Figure 5-58. Lighting Facts label.

or spaces with contact with insulation—important for products such as recessed downlights and cove lighting.

At the time of writing, there were roughly 450 products approved for the Lighting Facts label. Of these, about 45 percent were replacement lamps and 55 percent were luminaires and other products. Approved products can be viewed on the SSL Quality Advocates Initiative website at www.lighting-facts.com/products. ENERGY STAR-qualified products, Next Generation Luminaires and Lighting for Tomorrow award winners, and products that participated in DOE demonstration projects are highlighted.

Be warned, though, that rogue Lighting Facts labels are popping up, even among reputable product manufacturers. Each product is given an ID number for verification; use it.

Note that starting in mid-2011, packaging for all household screw-in lamps will be required to carry new labeling. Consumers are used to buying household lamps based on wattage, but as the market begins to shift to energy-efficient lamps, using wattage as the main criterion can be confusing. The Energy Independence and Security Act of 2007 required the Federal Trade Commission to consider new lamp labeling; the FTC announced new labeling requirements for front-end packaging that, as of the time of writing, were scheduled to become effective in mid-2011 (see **Figure 5-59**).

For the first time, the label on the front of the package will emphasize the lamp's brightness as measured in lumens, rather than a measurement of watts. The new front-of-package labels also will include the estimated yearly energy cost for the particular type of bulb. The new labels that focus on lumens are expected to help consumers make purchasing decisions as they transition to more energy-efficient lamp types.

The FTC labeling is mandatory, while the Lighting Facts label will remain voluntary. It is hard to say at this point how the two labels would be implemented together for LED replacement lamps, as they share some similar information and formatting.

Next Generation Luminaires Design Competition

Sponsored by DOE, IES and the International Association of Lighting Designers, the Next Generation Luminaires Design Competition was launched in 2008 to promote excellence in the design of energy-efficient LED luminaires for general lighting in commercial applications.

Here's how the program fits: Industry standards define how product performance should be tested and reported. The Lighting Facts label is a voluntary program for reporting this data in a consistent format for easy comparison. ENERGY STAR recognizes products that exceed minimum performance criteria. The NGL Competition takes this one step further, recognizing product design excellence using a broad list of criteria, with lighting designers and other professionals as judges. Among other criteria, a recognized winner needed to show innovation in terms of how LED

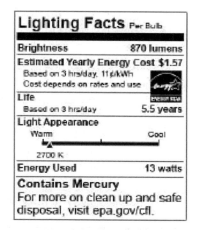

Figure 5-59. FTC's proposed Lighting Facts labeling.

New Back Label for Bulbs Containing Mercury

technology is integrated into new luminaire design, not just modifying existing products to accept LED emitters. Other factors such as good serviceability, thermal management, dimming capabilities, control of glare, mounting and others were criteria that aided a product being recognized as a specifiable product.

In its first year, the competition recognized 22 products from a total of 68 entries, or about one-third. In 2009, as the number of commercial LED lighting products on the market increased, the number of NGL entries nearly doubled—to 126, coming from 60 different lighting companies. Of these entries, 43 were chosen as "recognized" winners and four were chosen as "best in class," or a little more than one-third. This suggests that more product is entering the market but the level of risk is rising proportionately. LED luminaire manufacturers, in other words, are making more product sin more categories than ever before, but they are not proportionately doing a better job at designing them.

The product entries were grouped by application, which determined the specific performance categories and criteria used in their evaluation, and installed in environments closely matching the intended applications. A panel of 12 judges next independently evaluated each product based on their lighted performance and appearance, construction and photometric data. Judging criteria included evaluating overall construction, glare, color qualities, serviceability, ability to provide appropriate light levels and brightness, and aesthetics. Bonus points were available for luminaires that demonstrated no OFF-state power use, dark-sky friendly optics, adjustability and LM80 product depreciation (used to estimate service life) data.

The judges selected 47 products from the 126 entries as "recognized," meaning the panel would recommend these 47 products to other specifiers. To be recognized in this category, a product needed to achieve at least the judges' minimum evaluation in all performance categories and meet or exceed ENERGY STAR criteria where applicable. These products included accent lighting, wall washers, wall grazing luminaires, cove lights, undercabinet lights, refrigerated display lighting, downlights, general lighting, industrial lighting and an array of outdoor lighting.

From these products, the judges were asked to identify any that they considered "best in class"—standing out from the others and evaluated at the top of a majority of evaluation categories. These 2009 Best in Class winners included the CURVE task light by Finelite, Stile Styk wall washing luminaire by SPI Lighting (see Figure **5-60**), Evolve LED R150 Roadway luminaire by GE Lighting Solutions, and VizorLED parking garage luminaire by Philips Wide-Lite. These winners and all other recognized products can be viewed at www.ngldc.org.

Note the NGL completion does not address cost of equipment and long-term *in situ* performance, which is still largely an unknown at this time.

EMerge Standard Sets the Stage for LED Buildings

Today's workplaces are serviced by AC power but are often populated with DC-powered devices such as computers, phones and IT equipment. Lighting, traditionally an AC system, is predicted to become increasingly DC powered as LED and OLED lighting, DC-based fluorescent ballasting and buildingwide automatic lighting controls become more prominent in green buildings as well as mainstream workplace construction.

Using DC devices in a traditional AC power system, however, results in a lower level of utilization efficiency as AC must be converted to DC at the device level. The EMerge Alliance (www.emergealliance.org) was formed to address this problem by promoting a standard for a dual

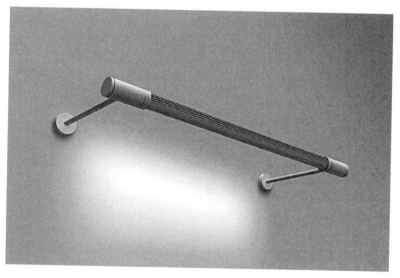

Figure 5-60. The judges in the 2009 Next Generation Luminaires Design Competition recognized the Stile Styk wall washer by SPI Lighting and three other products as "Best in Class." Photo courtesy of DOE.

power system that converts AC to DC power at the room level and then distributes it to DC devices in the space using a converter box and flexible DC wiring system (see **Figure 5-61**). This approach reduces energy losses associated with conversion at the device level, eliminates the need for separate components to transform AC to DC, and enables easy installation and movement of devices such as luminaires. The EMerge Standard also provides for the converter box to have an optional connection to directly receive power from on-site alternative power generation using sources such as wind and solar, which naturally produce DC.

In a typical system, Class 2 low-voltage wiring would form both a communication network and power grid in each primary space, with optional configurations using either powerline or wireless radio-frequency (RF) communication in addition to low-voltage power. Lighting devices such as luminaires and sensors would plug into the grid using snap connections to access power and communication. To the end-user, the system

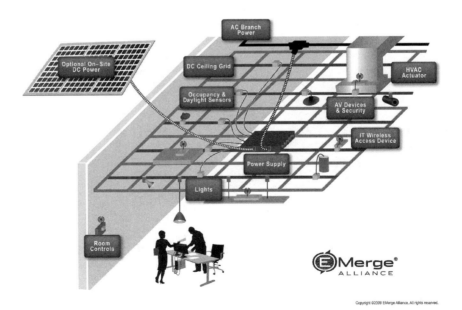

Figure 5-61. The dual power distribution system concept promoted by the EMerge Alliance involves converting AC to DC power at the room level and then feeding all DC-powered devices in the room, such as LED lighting and automatic controls, more efficiently. Image courtesy of the EMerge Alliance.

might look the same as today's lighting, but would no longer require each luminaire to be wired separately or to use a separate AC/DC power pack for power the control system. LED luminaires may also be as much as 15 percent more efficient than their AC counterparts because they will be driven directly by a DC power source.

Luminaire and control manufacturers, as well as other device manufacturers, are beginning to use the EMerge Standard, released in the fall of 2009, to build interoperable products that will provide green building solutions incorporating advanced, highly integrated and flexible lighting systems. At the time of writing, more than 50 manufacturers were participating in the Alliance, including lighting manufacturers such as Finelite, WattStopper, Zumtobel, Acuity Brands, Crestron, Lutron Electronics, Lucifer Lighting, Philips Electronics, Eden Park, Sensor Switch and Osram Sylvania. The first products are expected by the end of 2010 and will carry the EMerge registration mark, which means they have been evaluated by an authorized third-party lab to work with any other EMerge-registered products, and can be installed into any EMerge system. The most likely candidates are green buildings, owner-occupied offices and higher education.

Chapter 6

Lighting Maintenance

As lighting systems become increasingly complex,
maintenance requires more resources, expertise and competence.
—IES Recommended Practice for Planned Indoor Lighting Maintenance, p. 1

PLANNED LIGHTING MAINTENANCE

Lighting maintenance is essential because the performance of all lighting systems deteriorates over time, which can affect light levels and space appearance. Lighting components fail and must be replaced. Declining light levels, burnouts and color shift can affect the visual appearance of the space. And dirt and dust accumulation on luminaire and room surfaces, and deterioration of those surfaces themselves, will steadily reduce the amount of light output that reaches the workplane where it is needed. Maintenance can mitigate all of these problems.

There are two ways to conduct maintenance. One is to simply respond to problems as they occur. The second is to organize labor and resources to prevent problems before they occur, and maintain the system more efficiently. The second is called planned maintenance. Planned maintenance is more effective in that it offers a managed platform for continual lighting optimization.

Planned maintenance's benefits are based on three advantages. First, maintenance should have an objective of "keeping it working like new" versus "keeping it running." Second, all maintenance work should be planned out and scheduled. And third, maintenance work should be combined to economize on labor and resources. Virtually all planned maintenance programs include scheduled group relamping, luminaire cleaning and troubleshooting/inspection.

Compared to reactive maintenance, planned maintenance reduces operating costs, maximizes light levels, reduces waste, can improve safety and security, economizes on resources, and can improve space appearance.

Luminaire Cleaning

Cleaning the lighting system typically means washing or otherwise removing dirt and dust from lamps and luminaires (see **Figure 6-1**).

Cleaning lamps and luminaires maximizes luminaire light output by minimizing light loss due to absorption of light by accumulated dirt and dust. Luminaire cleaning increases the reflectance of interior luminaire surfaces, enables lenses to transmit more light, and helps maintain the luminaire's intended light distribution.

Luminaire cleaning can be economically combined with group relamping, inspection, repair and upgrades to energy-efficient lighting.

Group Relamping

Group relamping is the practice of replacing all lamps in a lighting system at scheduled intervals rather than individually as they fail. Benefits include lower maintenance costs, improved visual appearance, and minimized disruption to operations.

Why dispose of operating lamps? It is more economical overall to replace all lamps at fixed intervals rather than one at a time. It takes an estimated 50 hours

Figure 6-1. Luminaire cleaning raises light levels and helps maintain the luminaire's intended light distribution. Photo courtesy of Colorado Lighting, Inc.

to replace 100 lamps individually versus about 10 hours to group relamp them. Group relamping can reduce labor costs by as much as 80 percent. If (the labor cost of spot-replacing one lamp)—(the labor cost to group relamp it) > (the cost of a new lamp), then group relamping makes good economic sense.

But group relamping does not always boil down to labor costs. In the case of metal halide lamps, color shifting can occur near end of life, disrupting uniformity. Group relamping ensures that there is maximum lamp-to-lamp color uniformity. In the case of high-pressure sodium lamps, group relamping can reduce end-of-life cycling (automatic turning on and off of lamp) and subsequent wear and tear on the starter and ballast. In both cases, group relamping adds convenience when these lamps are in hard-to-reach luminaires. And in the case of all lighting systems, group relamping usually takes place after normal working hours, minimizing disruption to operations.

Group relamping can be economically combined with luminaire cleaning, inspection, repair and upgrades to energy-efficient lighting.

Figure 6-2. Group relamping economizes on labor and resources while improving the appearance of a space. Photo courtesy of Colorado Lighting, Inc.

Applications

Planned lighting maintenance can produce substantial benefits in most applications, but is especially beneficial in:

- installations where lamps operate on the same schedule, which makes the mortality rate easier to predict;

- installations with high, hard-to-reach mounting heights;

- metal halide lamp installations to minimize color shift and possibility of non-passive failure at end of life;

- environments with high levels of dirt and dust;

- buildings in which luminaires are being upgraded in a manner that will reduce light levels;

- fluorescent VHO lamp installations to compensate for high lamp lumen depreciation commonly experienced by these lamps as they get close to end of life; and

- indirect lighting installations, where luminaires are highly dependent on reflectances of luminaire and room surfaces to deliver light from the lamps to the workplane.

Maintenance as Energy-Saving Lighting Strategy

Lighting maintenance can play a part in a lighting upgrade project in three ways. First, proper maintenance can increase average light levels, potentially increasing design flexibility and energy savings in an upgrade. Second, it ensures that the new lamps continue to be installed over time without "snapback"—accidental reversion back to the old (usually cheaper) lamp type that fits the same socket. Third, maintenance provides another reason to replace older technology.

Planned maintenance can provide more flexibility to potentially leverage reductions in maintained light levels into additional energy savings. If planned maintenance raises maintained light levels by, say, 5 fc on the task in an open office, this provides additional "fat" for the lighting upgrade—light output that can be sacrificed to press energy savings.

Planned maintenance also can help prevent snapback; the maintenance policy should identify the lamp schedule to prevent gradual substitution of lamp types, which can jeopardize the intent of an original lighting design or upgrade. For example, suppose a lighting upgrade includes a switch from fluorescent T12 to 30W T8 lamps to save energy. Since the 30W and the standard 32W T8 lamps look the same and fit the same sockets, the maintenance department might begin replacing the 30W lamps with 32W lamps as they fail, disrupting uniformity and eroding energy savings. Similarly, the upgrade lamp and ballast schedule should be as simple and contain as few lamp and ballast types as possible.

Maintenance provides another reason to upgrade some lighting types beyond the energy savings argument. For example, probe-start metal halide lamps experience color shift over time, incandescent lamps provide short service life, high-pressure sodium systems cycle, etc. Replacing older systems with appropriate alternatives will not only save energy, but can also make maintenance easier. And when proposing the new alternatives, be sure to consider equipment such as extended-life lamps, programmed-start ballasts, parallel-circuit operation, luminaires with easily accessible features and mini-

Figure 6-3. Planned maintenance can increase light levels that can be leveraged into energy savings. Photo courtesy of Colorado Lighting, Inc.

mal components that can be easily removed for servicing, etc. (Also be sure to propose environmentally responsible disposal of replaced equipment.)

In the Spotlight: Designing for Maintenance

What the owner should ask:

- What are the designer's recommendations for maintenance so that the initial design can be sustained over time?
- What impacts would planned maintenance have on the lighting design versus reactive maintenance? What savings are possible in equipment and operating costs?
- Can the designer help the owner develop a written maintenance policy?
- What resources are available to safely recycle spent lamps?
- Does my organization have the capacity required to execute a planned maintenance program, or should we speak to a lighting management company that provides these services?

What the designer should provide:

- A written maintenance policy for the lighting system.
- Minimized number of lamp types used in the space, wherever practical, to reduce possibility of errors in lamp replacement.
- Luminaire locations that are sensitive to how a luminaire will be maintained (an incandescent pendant luminaire over an escalator, for example, might make a dramatic aesthetic impact but would be difficult and costly to maintain).
- An effective lighting design that meets the design intent for the space while minimizing operating and maintenance costs with proper maintenance methods.

The Planned Maintenance Policy

Being planned maintenance, it requires a written plan. The designer of the lighting system should provide this plan to the owner to ensure that the people maintaining the lighting system understand and support the intent of the people who designed it.

The written maintenance plan should include the design intent for the lighting and control system, schedule for the above maintenance operations, a complete listing of installed equipment, notification of special maintenance considerations, commissioning report for lighting controls, disposal recommendations and all operating and maintenance manuals.

The design intent is important so that the people who will be maintaining the lighting system can maintain the integrity of the design. Over time, maintenance operations can produce undesirable changes to the lighting design (e.g., causing accent lights to be aimed incorrectly, highlighting a blank wall next to a painting that was supposed to be highlighted), requiring maintenance personnel to understand the design intent helps ensure they will support it.

The maintenance plan should identify what planned maintenance operations will occur and on what schedule, including whether the lamps will be recycled or disposed in some other manner. The designer should communicate to the owner his or her maintenance assumptions, as this impacts light levels and other aspects of performance that the owner can expect.

A complete lighting equipment schedule includes a description of each type of luminaire for each space, without lamp and ballast catalog number and operating characteristics also identified. This will help ensure that when lamps and ballasts fail, they will be replaced with the exact same type, which helps avoid compromises to design integrity (e.g., mixing lamps with different color appearance in the same space) as well as "snapback" (e.g., replacement of energy-efficient lamps with lower-cost, less-efficient lamps that fit the same sockets).

The maintenance plan should include special maintenance considerations for the equipment—e.g., whether the space has high levels of dirt and dust in the air, if occupancy sensors are expected to frequently switch the lamps, whether extended-life lamps are used, or if any luminaires require periodic adjustment, such as aimable accent lighting.

All lighting controls should be commissioned and the results of this commissioning reported to the owner, with confirmation that any deficiencies are corrected and that the installed control system furnished to the owner achieves the control design intent without problems.

Finally, the owner should require the electrical contractor to turn over any operating and maintenance manuals provided for the installed lighting and control system.

To ensure the above is provided, the owner may request that the above documentation, and any training of maintenance personnel, be included in the specifications.

Maintenance Intervals

A maintenance interval is the point at which maintenance is conducted. If the group relamping interval is three years, for example, then every three years, group relamping will be performed. Finding the right interval for group relamping and luminaire cleaning requires analysis of a number of variables.

Luminaire cleaning interval. One way to determine the best luminaire cleaning interval is the point at which the cost of lost light output equals the cost of cleaning.

As dirt and dust accumulate on luminaire surfaces, these surfaces increasingly absorb light instead of reflecting it, and yet the owner is still paying to run the lamps. This means the owner is paying money to operate lamps from which they are getting less and less light. If the amount of this wasted money equals the cost of cleaning the luminaire, then the luminaire should be cleaned. If the owner waits even longer, the cost of doing nothing may exceed the cost of luminaire cleaning, which means the owner is essentially throwing money away.

More frequent cleanings may be scheduled, however. For example, the lighting system may be upgraded to an energy-saving lighting system (lower-wattage lamps and ballasts and/or fewer luminaires) that produces less light. By using frequent luminaire cleanings to optimize light output from the luminaire, more flexibility is gained in reducing the number of luminaires, lamps and/or system watts.

Group relamping interval. Looking at a typical fluorescent lamp's mortality curve, at 100 percent of the lamp's rated life, one-half of a large group of lamps can be expected to fail. One-half fail before, and one-half fail afterwards. The rate of failure, however, escalates sharply after 60-70 percent of rated life. As a rule of thumb, therefore, 60-70 percent is often considered a suitable group relamping interval. This can be assigned to a calendar interval (e.g., every three years).

From an economic point of view, the optimal group relamping interval can be calculated based on two factors: the cost of spot relamping and the cost of group relamping. As the failure rate accelerates, group relamping becomes less expensive than spot relamping. Group relamping is more attractive when lamps are cheap and spot replacement labor is expensive. Computer software is available that can help calculate the optimal group relamping interval.

Group relamping intervals may not always be calculated on the lamp's mortality rate, however. In the case of some lamp types, the lamp

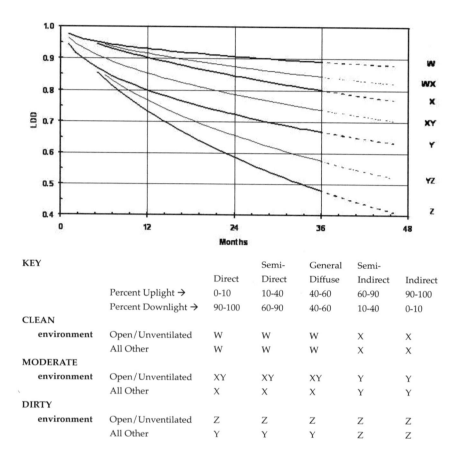

KEY		Direct	Semi-Direct	General Diffuse	Semi-Indirect	Indirect
	Percent Uplight →	0-10	10-40	40-60	60-90	90-100
	Percent Downlight →	90-100	60-90	40-60	10-40	0-10
CLEAN						
environment	Open/Unventilated	W	W	W	X	X
	All Other	W	W	W	X	X
MODERATE						
environment	Open/Unventilated	XY	XY	XY	Y	Y
	All Other	X	X	X	Y	Y
DIRTY						
environment	Open/Unventilated	Z	Z	Z	Z	Z
	All Other	Y	Y	Y	Z	Z

Figure 6-4. Luminaire dirt depreciation (LDD) curves for various fluorescent indoor luminaires. The left axis shows percentage of light escaping the luminaire; the bottom axis shows time in months. As can be seen, light output from a luminaire steadily decreases over time based on a number of factors, including the luminaire type and how much dirt is present in the space. The bottom line is that the luminaire's owner ends up paying for light that never leaves the luminaire—more and more over time. By cleaning the luminaire, more usable light will be produced, in addition to other benefits. Image courtesy of NALMCO.

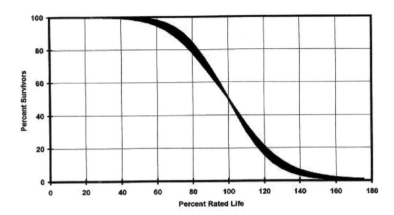

Figure 6-5. Typical mortality curve for fluorescent lamps. Large groups of lamps tend to follow this curve. Since most lamps start failing after 60-70 percent of lamp life, it typically makes economic sense to replace all of the lamps in the system at the same time instead of replacing subsequent failures individually. This can produce significant labor savings. Image courtesy of the interNational Association of Lighting Management Companies (NALMCO).

does not fail, but produces so little output that it is no longer economical to continue operating. An example is a metal halide or mercury vapor lamp that continues to light, but has become extremely dim due to lumen depreciation. Useful life can be defined several different ways, therefore. One way is if the lamp has depreciated to 50 percent of its initial light output (70 percent if SSL). Another is if the lamp has undergone significant color shift. A final way is if the lamp is cycling or experiencing some other defect. In all cases, lamps that have passed their useful life should be replaced.

EQUIPMENT SELECTION

The choice of lamps, ballasts and luminaires all affect lighting maintenance. Indirect luminaires, for example, can be desirable for lighting quality, but have exposed reflectors and are more sensitive to dirt and dust buildup. Below are several considerations for optimizing maintenance during specification; ask your system designer or distributor for guidance on product selection.

- Consider long-life light sources, particularly in hard-to-reach luminaires.

- LED light sources in exit signs and other applications are considered long-life light sources, but note that they may experience a dramatic loss of light output prior to actual failure—they should be replaced when light output declines to a point where the product is no longer usable, not when it fails. With LEDs used as a general light source, this is when the LEDs experience a loss of light output of about 30 percent.

- Consider ceramic metal halide lamps instead of standard metal halide lamps not just for efficiency gains, but to eliminate color shift and improve lamp-to-lamp color consistency.

- Consider programmed-start ballasts that optimize fluorescent lamp life in applications where the lamps are frequently switched, such as occupancy sensor installations, as frequently switching can reduce lamp life.

- Avoid incandescent lamps wherever possible.

- Consider ballasts that operate lamps on a parallel circuit instead of a series circuit, so that if one lamp operated by the ballast fails, the other will continue to operate normally.

- Consider luminaires with easily accessible features, optical systems that suit the environment, minimal components that be easily removed for servicing, and, where required, robust gaskets and/or seals.

Proper maintenance should be a leading consideration during equipment selection and planning a lighting design or lighting upgrade. It can ensure long-term design integrity while optimizing cost savings.

LIGHTING DISPOSAL

Planned lighting maintenance and lighting upgrade operations can result in wholesale replacements of lamps and ballasts. Once lamps and ballasts are destined for disposal, they become "waste" and state and Federal regulations may come into effect.

In the Spotlight: Extended-life Fluorescent Lamps

As the T12 lamp reaches the end of its product cycle, T5 and T8 are natural technology replacements. T8 lamps are now into their second and third generation with added features, including longer service life, which can reduce lamp replacement and disposal costs.

Extended-life lamps, available with instant-start or programmed-start ballasting, can provide a rated service life up to 40,000 hours at 12 hours per start on instant-start ballasts—the most popular electronic ballast type—and up to 46,000 hours at 12 hours per start on programmed-start ballasts which are ideally suited for maintenance-sensitive and frequently switched (i.e., occupancy sensor) applications. The gain in service life is due to a heavier cathode that can carry more emission material or the fill gas can be modified or its pressure raised.

The industry average life for a 4-ft. T8 lamp is 20,000 hours at three hours per start and 24,000 hours at 12 hours per start on an instant-start ballast, and 24,000 hours at three hours per start and 30,000 hours at 12 hours per start on a programmed-start ballast.

Today's extended-life fluorescent lamps include premium fluorescent lamps offering life ratings of 24,000 to 36,000 hours at three hours per start and 30,000 to 40,000 hours at 12 hours per start on an instant-start ballast, and 24,000 to 40,000 hours at three hours per start and 36,000 to 46,000 hours at 12 hours per start on a programmed-start ballast. Some of these lamps are premium versions of standard lamps, combining the benefit of longer life with other high-end performance features such as additional energy savings, low mercury, RoHS compliance and in some cases an extended lamp warranty. The lamps at the high end of the range of rated life may be recognizable via a suffix at the end of the lamp's nomenclature, such as XL, XLL (Philips Extra Long Life) and SXL (GE Super X-tra Life).

Table 6-1. Rated life (hours) for industry-average 4-ft. T8 lamps and extended-life T8 lamps on an instant- and programmed-start ballast at three and 12 hours per start.

	Instant Start Ballast		Programmed Start Ballast	
	3 hrs/start	12 hrs/start	3 hrs/start	12 hrs/start
Industry Average (32W T8)	20,000	24,000	24,000	30,000
"XL" Lamps (25-32W T8)	21,000-36,000	30,000-40,000	30,000-40,000	36,000-46,000
% Greater Rated Life	5-80%	25-67%	25-67%	20-53%

> The basic cost of light from a single lamp is 4-8-88: About four percent is the cost of the lamp, eight percent is the labor to install the lamp, and 88 percent is the energy required to operate it over its life. The labor cost, in other words, is often twice the initial cost of the lamp. If a $2.00 lamp lasts three years, the cost is $0.67 per year; if a $2.40 extended-life lamp lasts five years, the cost is less than $0.50 per year, coupled with the benefits of a longer maintenance cycle and reduced disposal.

Lamps

Fluorescent, high-pressure sodium, metal halide, mercury vapor and neon lamps that contain mercury may be considered hazardous waste when they have failed and are ready for disposal. The U.S. Environmental Protection Agency (EPA) defines a hazardous waste lamp as a lamp that is characteristically hazardous, meaning it fails EPA's Toxicity Characteristic Leaching Procedure (TCLP) for mercury.

Disposing of hazardous waste has significant regulatory requirements. In 2000, EPA changed the rules for mercury-containing lamps to allow them to be classified as universal waste, with fewer regulatory requirements, if they are recycled. A number of companies in the U.S. provide lamp and ballast recycling services.

Options for managing the disposal of mercury-containing lamps according to Federal requirements, therefore, include treating them as hazardous waste, treating them as universal waste (recycling), or to use a type of lamp that is not characteristically hazardous. Recycling typically costs more than hazardous waste landfill costs, but with less regulation and paperwork, and lower storage, collection and transportation fees. The owner also benefits from reduced liability.

In recent years, however, the major lamp manufacturers have voluntarily reduced the amount of mercury in their lamps and made other changes in the operating design of their fluorescent lamps so that they pass the TCLP test and therefore can potentially be disposed of in a municipal landfill, depending on state laws. It is important to note that while the Federal government mandates minimum requirements, state governments may create stricter requirements. California, Connecticut, Florida, Maine, Minnesota, Rhode Island and Vermont, for example, have banned all mercury-containing lamps from solid waste landfills. And Pennsylvania has eliminated the small-quantity exemption. These regulations change over time, so the reader should investigate the latest regulations applicable to their projects.

General guidelines for fluorescent lamp disposal, published on the EPA's web site, are shown below. For a list of lamp recyclers and links to lamp disposal regulations that can vary from state to state, visit www. lamprecycle.org, a web site produced by the National Electrical Manufacturers Association.

Disposal methods
1. Place waste lamps in the box in which replacement lamps arrived, or in special cartons provided by the lamp recycler.
2. Store lamps in a safe place to avoid breakage, marking the area appropriately to prevent others from accidentally throwing trash in it.
3. Separate and put broken lamps in a heavy plastic bag placed inside a rigid container. If you cannot locate a lamp recycler who will accept them, treat broken lamps as hazardous waste.
4. Do not place broken fluorescent lamps in metal receptacles. Metal boxes will absorb mercury and become hazardous waste containers.

Collection and Transportation
1. Most lamp recyclers offer transportation services. Waste lamp generators may also contract with a solid or hazardous waste transporter to take lamps to a recycler, or safely transport their lamps themselves.
2. Lamp generators may collect waste lamps from several locations and store them in a central facility to ease transport and recycling.
3. Transportation to another state may require use of a transporter licensed in that state and compliance with that state's hazardous waste transportation laws.

PCB Ballasts
PCBs are a hazardous waste regulated by the TSCA Section of EPA regulations and were once used in the capacitor of rapid start magnetic ballasts. The U.S. banned the manufacture and distribution of PCB-containing ballasts in 1978, but they are still found in older luminaires. When a PCB ballast is encountered, it must be disposed using strong precautions; EPA requires that it eventually be disposed in a Federally approved incinerator. If the ballast is not leaking PCB-containing fluid, often the surest and simplest method of disposal is to bring in a ballast recycler. If the ballast is leaking, then it must be treated as hazardous waste

and incinerated. Whether the ballast is leaking or not, a qualified disposal contractor should handle its disposal.

COMMISSIONING LIGHTING AND CONTROL SYSTEMS

Commissioning is a quality assurance process used to ensure that an installed building system performs according to the design intent and owner's operational needs, and also achieves user acceptance.

According to the U.S. Green Building Council (USGBC), "Benefits of commissioning include reduced energy use, lower operating costs, fewer contractor callbacks, better building documentation, improved occupancy productivity and verification that the systems perform in accordance with the owner's project requirements."

The USGBC's LEED 2009 green building rating system requires commissioning of building control systems as a prerequisite and encourages advanced commissioning with two LEED points. LEED identifies specific requirements, but a more detailed commissioning process may be found in ASHRAE Guideline 0-2005.

In larger projects, third-party commissioning is recommended. LEED 2009 for New Construction and Major Renovations, in fact, recognizes it for projects 50,000 sq. ft. and up, with the commissioning authority reporting directly to the owner. In smaller projects, the commissioning authority may be the person who designed the given system. A good specification will not only require commissioning but identify parties and responsibilities. As commissioning may require functional testing of systems using high voltages, the electrical contractor is involved in the process.

When one talks about commissioning for lighting, they are usually talking about lighting control systems such as automatic shutoff or daylight harvesting. If controls are misapplied, incorrectly installed or do not perform according to design intent, users may bypass them and the owner may remove them entirely. The more complex or application-sensitive the control system is, the more commissioning can distinguish whether it will succeed or fail. The basic process involves verifying installation, performing operational checks, and training owner staff on proper use of the system. Deficiencies are recorded and promptly corrected.

The commissioning process may also be applied to the controlled lighting equipment and even daylighting systems, such as light shelves, with similar benefit, however. While not required by LEED, commissioning these items can achieve the same benefits. The IES formed a committee to develop a recommended practice for commissioning lighting systems, written with the assistance of this book's author, that is expected to be published within the next year. In the meantime, the Lighting Controls Association has published a course on commissioning at www. aboutlightingcontrols.org, with worksheets that can be used to perform functional testing of time sweep, occupancy sensor, daylight harvesting and architectural dimming control systems.

As a critical player in the commissioning process, electrical contractors should be familiar with the complete process and be prepared to work with designers and commissioning agents to satisfy project requirements. Further, contractors that gain experience and expertise in this area may be able to leverage these skills as a competitive advantage on LEED projects and potential business opportunities retro-commissioning existing lighting and control installations.

Design Intent

Commissioning begins with clearly expressed owner project requirements and design intent, also called the basis of design. The project requirements may include the desired level of quality, reliability, automation, flexibility and ease of use and maintenance for the given system. The owner may also wish to identify energy efficiency targets and preferred technologies and manufacturers. The design intent is based on these requirements and includes performance criteria, applicable codes and standards, functionality and, for lighting controls, a sequence of operations for each control point.

System Activation

The first real step is to activate the lighting and control system. For the lighting system, this involves energizing the luminaires and operating them for a period of time without dimming to stabilize them. Aimable luminaires may be pre-aimed by the contractor per the drawings. For the controls, this step typically would entail control manufacturer representatives working with the electrical contractor to calibrate and adjust the controls to match the specifications and site conditions.

Installation Verification

After activation, the commissioning authority can verify installa-

Figure 6-6. A typical performance testing procedure for a low-voltage relay-based time-sweep control system, installed to provide simple ON/OFF automation of large loads, includes: First, ensure that the lights in the control zone are turned OFF at the scheduled time. Second, confirm that the lighting sweep is preceded by a warning to the occupants, such as blinking or audible signal. Third, make sure the override only turns the override zone ON. Fourth, ensure that after the schedule is overridden, the lights will subsequently sweep OFF after the specified period of time. Fifth, if there is a key-activated whole floor switch (for cleaning staff, etc.), confirm that it is also swept OFF. And sixth, if there is split wiring in the space (i.e., multilevel switching), test the controls to ensure the different zones are properly sequenced. Photo courtesy of Schneider Electric.

tion. This involves ensuring that the equipment nameplate data matches the approved submittals. Luminaires and special mounting hardware such as stems, etc. should be inspected to ensure they are properly located and aligned, clean and undamaged, contain the correct lamps and ballasts, have no exposed wires, and have the specified characteristics—color, louvers, cones, shields, filters, lenses, etc.

Here are a few things to look for during this step: There should be no light leaks. Pendants should be mounted at the specified height and properly aligned. Square luminaires should be installed parallel to walls or tiles or other reference point. Joints should be aligned. Holes in ceilings and walls should be properly sized to be covered by canopy or luminaire flange or trim. Canopies should be flush as required. Wall washers should aim at the correct wall. In addition, there should be no obviously defective lamps or ballasts identifiable as such by excessive ballast noise, lamp flicker, burnouts, lamp end blackening.

The commissioning authority may need to verify appropriate reflectances on room surfaces if values are included in the performance criteria in the basis of design and approved design. Reflectance expresses the amount of light a given surface reflects instead of absorbs. Measure the reflectance of ceiling, walls and floors as needed to verify appropriate values based on the specification. Reflectance may be simply measured by taking a footcandle reading holding a light meter flat against the wall (with the sensor facing out) and then a second reading with the meter held one to two feet from the wall, with the sensor facing the wall. The second reading divided by the first is reflectance, expressed as a percent.

Similarly, the commissioning authority will verify that the right control equipment was installed, that it was installed properly, and that it provides the desired functionality. This process may be expedited by performing installation verification in a sampling instead of every single control zone. Specific verification and testing procedures may vary by type of control system, and often include confirming the system delivers performance characteristics specified by energy codes, such as maximum size of override zones in a time-sweep automatic shutoff system. Further, the authority should look for any site conditions that are likely to become deficiencies in meeting the design intent.

At the conclusion of the installation verification phase of commissioning, any damaged or defective components should be promptly replaced and any other deficiencies discovered should be quickly corrected so that functional testing can begin.

Performance Testing

This step, also called field commissioning, involves ensuring the installed system meets the design intent under typical operating conditions. First, the building must be made ready. The building should be enclosed and all windows and outside doors installed and closed; all furniture and ceiling tiles should be installed, and permanent power supplied to all luminaires. The contract should also ensure that all equipment needed for commissioning, such as ladders or lift equipment, is on-site ready or use.

The lighting system should be visually inspected during operation to ensure the lamps have an appropriate color tone and the luminaires are producing an appropriate distribution pattern. Specific tests should be performed as required to confirm desired initial horizontal and vertical light levels, desired uniformity of light intensity on the task plane, and, in the case of emergency lighting, compliance with specific codes. Any footcandle measurements used to determine light levels or reflectances should be taken with a properly calibrated meter, and using appropriate procedures. Aimable luminaires should receive final aiming (which may require ladders, gloves, walkie-talkies, tools, etc.) and be locked into position.

The functional testing regime for the lighting controls will depend on the type of lighting control system installed. It may involve programming of time-of-day scheduling, overrides, preset dimming scenes and user PC controls. Sensors may require recalibration to achieve desired performance; note that some controls are designed to be self-calibrating, while others can be more conveniently calibrated set up remotely using handheld remotes or a computer with an appropriate connection. As with installation verification, the process may be expedited by testing a sampling instead of every control zone. Points of interaction between different control devices and systems, and between the lighting control system and other building systems such as a building automation, security and A/V, may require additional testing.

Owner Notification

The last step in commissioning is owner notification. Facility operators and occupants should be explained the functionality of the lighting and control system, including local control capability such as overrides and personal dimming, programming and so on.

As part of this, maintenance personnel should receive live (may also be prerecorded) training on the proper operation and maintenance of the lighting and control system, including demonstration. All documentation and instructions for ongoing operations and maintenance should be turned over, including an operations and maintenance manual submitted by the contractor, latest shop drawings, installation and troubleshooting guides, calibration settings, panel schedules, warranties and distributors for each component. Finally, the turnover phase may also include delivery of spares, which may include 10 percent overage on lamps, five percent overage on ballasts, 10 percent overage on lenses and baffles, and any other spares required by the specifications.

Even after the project is turned over to the owner, commissioning does not end. LEED 2009's EA Credit 3: Enhanced Commissioning, valued at two LEED points, requires review of the operation of the building with facility operating staff and occupants within 10 months after substantial completion, with a plan to resolve any outstanding commissioning issues.

Appendix

Resources

Personally, I would hate to live in a world where
lighting design was reduced to a paint-by-numbers exercise.
—Leslie M. North, lighting consultant

ENERGY CODES

Commercial building energy codes vary by state and even local jurisdiction. Despite convergence in the major energy standards—ASHRAE's 90.1 and ICC's International Energy Conservation Code (IECC)—the nation is a patchwork of codes, some based on ASHRAE 90.1 (1989, 1999, 2001, 2004, 2007), with or without amendments; some based on IECC (2003, 2006, 2009), with or without amendments; and others completely homegrown, such as California's strict Title 24 energy code.

To keep up with codes and learn more about the code in fact in your jurisdiction, two resources may help. First is DOE's www.energycodes.gov, which provides resources about energy codes, including compliance software. Second is the Building Codes Assistance Project at www.bcap-energy.org.

LIGHT + DESIGN

Light + Design: A Guide to Designing Quality Lighting for People and Buildings, edited by this book's author and published by the Illuminating Engineering Society of North America, provides a practical introduction to the world of lighting design. Supplementing the Guide are questions that provide a checklist of critical issues plus an applications guide. Numerous sketches and photos, glossary and end notes are included. This 192-page publication, published in hardcover in an 8.5 x 11 size, is available for $65.50 to IES members and $95.00 to non-members.

For more information, visit www.ies.org.

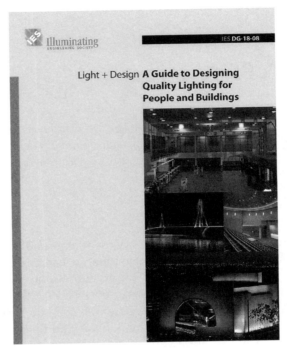

Figure A-1. Light + Design provides a practical introduction to the world of lighting design.

COMMERCIAL BUILDINGS TAX DEDUCTION

The Commercial Buildings Deduction, recently extended to December 31, 2013, enables commercial building owners to claim an accelerated tax deduction of up to $0.60/sq.ft. for investing in energy-efficient interior lighting. The Interim Lighting Rule offers a simple path to qualification, as long as the project includes required lighting controls and meets IES light levels. Projects must be certified by qualified individuals. The National Electrical Manufacturers Association (NEMA) recently updated its Commercial Buildings Tax Deduction website, written by this book's author, to provide highly detailed guidance about how to qualify for this incentive.

Find it here: www.lightingtaxdeduction.org.

LAMP RECYCLING

NEMA's lamprecycle.org website has provides a one-stop source of information about recycling mercury-containing lamps. The latest version of the site allows visitors to find compact fluorescent lamp drop-off locations near them with the help of Earth 911.com.

ARCHITECTURAL LIGHTING MAGAZINE

Architectural Lighting Magazine has covered cutting-edge lighting design projects and products for more than 20 years. Learn more at www.archlighting.com.

LIGHTING BLOG

LightNOW, launched at Lightsearch.com in 2001 and produced by this book's author, is a blog where visitors can read daily postings to learn about the latest happenings in the lighting and related industries. For those who want their information in a traditional newsletter format, the month's posts are aggregated into a newsletter and emailed to more than 14,000 architects, designers and other construction professionals around the world.

Check it out at www.lightnowblog.com.

CONTROLS AND CODE

The Lighting Controls Association has published a free online course on commercial energy codes as part of its Education Express program. EE203: Lighting and Commercial Energy Codes, written by this book's author, outlines the lighting requirements of the International Energy Conservation Code (2003, 2006, 2009) and ASHRAE 90.1 (1999, 2001, 2004, 2007), with special focus on making the right lighting controls choices for compliance. Process maps outlining each step towards compliance provides a graphical learning experience and enables students to visualize the process. Get educated at www.aboutlightingcontrols.org.

ARCHITECTURAL SSL MAGAZINE

Architectural SSL is a quarterly journal about the application of solid-state lighting in the built environment. Now you can get additional information at SSL Interactive, a blog providing product and project reviews, technical data and other stories. The blog, and access to articles in past issues, can be read at www.architecturalssl.com.

LIGHTING DIRECTORY

Since 1996, Lightsearch.com has served the lighting industry as its most popular directory. To access a listing in the site's database of more than 5,500 manufacturers, simply type in the name of the company in the search box. Alternately, one can search by product category from among 11,000 categories of luminaires, lamps, ballasts, controls, components, software, LEDs, fiber-optic and daylighting products. Site users can also access a directory of design firms and get industry news. Check it out at www.lightsearch.com.

LEARNING TO SEE

Howard Brandston's book *Learning to See: A Matter of Light*, published by the Illuminating Engineering Society of North America, is the legendary lighting designer's gift to the design community, a playful but incisive distillation of more than 50 years designing and teach lighting. This is not a technical publication, but a treatise on the art and science of lighting, and defines a creative process that applies to any type of design. The book, edited for publication by this book's author, is available at www.ies.org/store.

Figure A-2. Howard Brandston's book is a treatise on the art and science of lighting.

LIGHTING CLASSES

The Lighting Education Institute is exclusively dedicated to professional lighting education. Founded by Craig A. Bernecker, who developed and directed the lighting program at Penn State University, The Lighting Education Institute's goal is to bring that same quality of education directly to the building industry to enhance the abilities of practicing professionals and to better prepare those who are entering the profession. The Lighting Education Institute provides regular classes in which individuals can enroll, but also tailors lighting education programs to the specific needs of a group or company. For more information, visit www.lightingeducation.com.

ELECTRICITY PRICES

If you do not know what electricity prices are for a given utility, you can quickly look up average prices in the state, courtesy of the Energy Information Administration, a division of DOE. The average retail price of electricity is provided by end-use sector—commercial, industrial, residential and transportation—and by state, and contrasted against the previous year for a sense of where prices are moving and by how much. Links provide access to historical reports and other useful information about energy consumption, such as lighting's share of building energy use. Get the data at http://www.eia.doe.gov/cneaf/electricity/epm/table5_6_a.html.

DAYLIGHTING STUDIES

Heschong Mahone Group conducts daylighting, code, energy benchmarking and technical research for government, utilities, environmental groups and other organizations. On behalf of PG&E and the California Energy Commission, Heschong Mahone conducted a series of studies documenting the impact of daylighting and windows on office worker satisfaction, retail sales and student performance in schools. In brief: Daylighting can make a significant positive impact on all of these areas. Download the studies at www.h-m-g.com.

ENERGY STAR

ENERGY STAR criteria have been developed for a wide range of electrical devices used in homes, from compact fluorescent lamps to washers and dryers. In 2008, the first criteria for LED luminaires became effective, with the first products entering the market in 2009.

ENERGY STAR criteria now cover recessed downlights, undercabinet kitchen lights, shelf-mounted display and task lights, portable desk lamps, cove lighting, ceiling-mounted, surface-mounted with directional heads, and outdoor wall-mounted porch, step, pathway and pole/arm-mounted decorative lights. LED luminaires labeled with the ENERGY STAR mark not only meet a certain threshold for efficiency, but also performance. Purchasers gain some confidence that the product will perform as well or better than similar product using traditional light sources.

DOE has also developed criteria for integral LED replacement lamps, published in December 2009 and to become effective August 31, 2010.

See a list of ENERGY STAR-labeled products at www.energystar. gov.

NEXT GENERATION LUMINAIRES COMPETITION

The Next Generation Luminaires Solid State Lighting Design Competition was created to recognize and promote excellence in LED commercial lighting product design.

After demonstrating that it meets ENERGY STAR criteria that may be applicable, a given product is put through a tough review by a panel judges including representatives of the International Association of Lighting Designers, Illuminating Engineering Society and DOE. They evaluate the products' lighting performance, color characteristics, construction, serviceability and submitted photometric data.

ENERGY STAR is a minimum threshold of performance, and does not cover everything that makes a lighting product worth specifying. This competition identifies truly outstanding products as judged by leading lighting designers. The competition results enable lighting decision makers to vet products as being ready to specify or not.

See the winning products at www.ngldc.org.

Figure A-3. The Next Generation Luminaires Design Competition recognizes and promotes excellence in LED commercial lighting product design.

DALI BY DESIGN

DALIbyDesign, built by this book's author on behalf of RNM Engineering, is an impartial website dedicated to educating lighting practitioners, building owners and other professionals about digital lighting control—specifically, digital control systems based on the Digital Addressable Lighting Interface (DALI). Find it at www.dalibydesign.us.

LIGHTING DESIGN LAB

The Lighting Design, based in Seattle, is dedicated to transforming lighting in the Northwest by promoting quality lighting design and energy-efficient technologies. The Lab accomplishes this through education and training, consultations, technical assistance and demonstrations.

Check out this strong resource at www.lightingdesignlab.com.

SSL QUALITY ADVOCATES

The Quality Advocates Initiative was created by the Next Generation Lighting Industry Alliance and DOE to encourage LED product manufacturers to voluntarily follow certain guidelines and labeling to report performance claims.

The Lighting Facts label promotes accurate and consistent reporting of product performance. Lighting decision makers benefit by knowing with greater certainty how a given product will perform, more easily comparing different products, and being given more data about performance that serves as a reminder to consider the product's entire story, not just its wattage.

Products are tested using the IES LM-79 standard, and the testing data verified as part of registration. Once registration is complete, the product can carry the Lighting Facts label, which summarizes performance data covering light output, input watts, efficacy, color appearance and color rendering.

Be warned, though, that rogue Lighting Facts labels are popping up, even among reputable product manufacturers. Each product is given an ID number for verification; use it.

Learn more at www.lightingfacts.com.

ADVANCED LIGHTING GUIDELINES

The Advanced Lighting Guidelines is published by the New Buildings Institute to provide lighting practitioners with a practical guide on how to achieve good lighting design that maximizes both quality and efficiency.

The latest edition, the 2010 edition, was in production as a web-based publication at the time of writing. This book's author contributed the chapter on lighting controls with Dorene Maniccea of WattStopper.

Learn more at www.newbuildings.org.

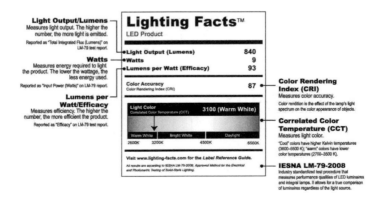

Figure A-4. The Lighting Facts Label.

COMMERCIAL LIGHTING SOLUTIONS

The Commercial Lighting Solutions program, sponsored by DOE, is a webtool providing customizable lighting templates designed to generate more than 30 percent lighting energy savings compared to ASHRAE 90.1-2004, without sacrificing lighting quality. It is specifically designed for people who make lighting decisions but are not necessarily lighting experts. Contractors, designers, distributors and owners can use these templates to achieve the latest, energy-efficient solutions—while providing good lighting.

CLS for Retail launched at Lightfair 2009. CLS for Office was subsequently fast-tracked to support large public spending projects, and was launched at Lightfair 2010. This book's author assisted in the development of the control templates for the Office version.

The CLS webtool is available free to the public and can be accessed at www.lightingsolutions.energy.gov.

Figure A-5. DOE's CLS webtool provides customizable lighting templates designed to generate more than 30 percent lighting energy savings compared to ASHRAE 90.1-2004 for retail and office buildings.

ADVANCED ENERGY DESIGN GUIDES

ASHRAE, in collaboration with government and other industry organizations, developed a series of Advanced Energy Design Guides that provide a prescriptive path for beating ASHRAE 90.1-1999, the national energy standard (prior to 2011), by 30 percent in small office buildings, small retail buildings, K-12 school buildings and other building types.

To download any of ASHRAE's Advanced Energy Design Guides free, visit ashrae.org.

INTERNATIONAL DARK SKY ASSOCIATION

The International Dark-Sky Association is a nonprofit organization dedicated to protecting and preserving the nighttime environment through quality outdoor lighting.

Learn more at www.darksky.org.

Index